Divided Natures

French Contributions to Political Ecology

Kerry H. Whiteside

The MIT Press
Cambridge, Massachusetts
London, England

Set in Sabon by The MIT Press.
Printed (on recycled paper) and bound in the United States of America.

Library of Congress Cataloging-in-Publication Data

Whiteside, Kerry H., 1953–
Divided natures : French contributions to political ecology / Kerry H. Whiteside.
p. cm.
Includes bibliographical references.
ISBN 0-262-23221-9 (hc. : alk. paper) — ISBN 0-262-73147-9 (pbk. : alk. paper)
1. Political ecology—France. 2. Political ecology. I. Title.
JA75.8 .W55   2002
320.5—dc21   2001044443

Contents

dedicated to my parents

Acknowledgements

I want to express my thanks to the following individuals and publishers for giving their authorization to reprint selected passages: Alain de Benoist, editor of *Krisis*, for Serres 1993; *Nouvel Observateur* for Serres 1992c; Bruno Latour for Latour 1987a; *Theory, Culture & Society* for Moscovici 1990.

Portions of the book originated in my published articles, which I have since revised and reorganized. I want to thank the following journals for allowing me to draw upon this material: *French Politics, Culture, and Society* for Whiteside 1995 (in chapters 4 and 6); *Contemporary French Civilization* for Whiteside 1997a (in chapters 1 and 6); Frank Cass Publishers for Whiteside 1997b (in chapter 6); Guilford Press for Whiteside 1996 (primarily in chapter 8).

It would be difficult, and far less stimulating, to write on French thought without doing research in France. For trips to France, I have been fortunate in receiving financial support from Franklin and Marshall College and from the Institute for European Studies. Both have my gratitude. Their support increased my opportunities to meet many participants in and students of the French environmental movement. I have benefited greatly from conversations with Alain Lipietz, Jean-Paul Deléage, Guillaume Sainteny, Daniel Boy, Yves Cochet, Jean-Louis Vidal, Dominique Allan Michaud, Pierre Lascoumes, Denis Duclos, and Pierre Radanne. My thanks go to all of them, as well as to various workers in the national headquarters of Les Verts, who graciously made time to aid a foreign researcher in their midst.

Scrutiny by colleagues in political theory and comparative politics has also made the book better than it otherwise would have been. I have depended on daily conversations with Dean Hammer for insight on everything from trends in political theory to word choice. Others who have encouraged me in my work and commented constructively on this project include Lew and Sandy Hinchman, Jane Bennett, Harlan Wilson, Tad Shull, Peter Cannavo, and Mary Bellhouse.

It has been a great pleasure to work with Clay Morgan of The MIT Press. His editorial advice greatly eased movement toward completion of the project.

Divided Natures

Introduction

Could it be that we have unfairly neglected French contributions to green theory because of words written more than 350 years ago?[1] It was a sixteenth-century Frenchman who, in the opinion of many green thinkers, penned the most notorious line in the history of Western philosophy. René Descartes (1637: 40) proposed that we "make ourselves masters and possessors of nature"[2] by subjecting our material environment first to rational analysis and then to technological control. Fritjof Capra (1982: 61) speaks for many ecologists when he charges that the Cartesian view of the universe "provided a 'scientific' sanction for the manipulation and exploitation of nature that has become typical of Western culture."[3]

If Descartes's views are felt throughout Western culture, his influence has been even more pronounced in France. In France from the seventeenth century on, according to H. Stuart Hughes (1966: 4), "Cartesianism suffused the intellectual atmosphere so thoroughly that much of the time it went unnoticed"; Descartes was France's "official philosopher." French observers themselves acknowledge Descartes's impact on the aesthetics of their physical surroundings. In France, Roger Cans remarks (1992: 218), people "always favor a nature that has been domesticated, subdued, divided up." Even Jean Jacob, the author of the most comprehensive French-language study of ecological thought, calls his country "the land of artifice" and uses this notion to explain how hard the ecology movement has had to struggle to gain credit there (Jacob 1999: 310).[4] Anyone who has contemplated the regimentally aligned trees and the geometrically sculpted greenery of the Versailles gardens may find it difficult to suppress the suspicion that ambient Cartesianism makes France barren ground for the cultivation of environmental concern.[5]

That suspicion is both unjustified and misleading. It is unjustified because since the 1950s the French have generated an abundant and original literature of environmental political thought. Comparable in its intellectual sweep to the range of green theory available in English, French ecologism deserves attention that it has rarely gotten. That is why I began a book that would serve as a bridge between versions of green political theory in two linguistic communities. French green thinkers such as Edgar Morin, Michel Serres, Bruno Latour, Alain Lipietz, and Denis Duclos merit a place in the otherwise robust and international-minded discussions of the aims of environmental politics.

As I proceeded in my study, however, it became clear that more was at stake than simply widening the field ecological discourse. Neglect of French ecologism, I came to conclude, misleads us by skewing understandings of environmental thought in general. Omitting the French from general accounts of ecologism reinforces the impression that debate over the roles of "humanity" and "nature" in instituting environmental values is the central controversy in the field of green political theory.

This debate gets played out through a key distinction that finds its way into almost every philosophical discussion of ecologism written in English. "Anthropocentric" ecologists contend that whatever reasons we have to protect our nonhuman surroundings derive ultimately from their role in fulfilling human interests and values. Calling the contrasting position "ecocentric," Robyn Eckersley (1992: 26) defends an ecologism that recognizes, in addition to human values, "the moral standing of the nonhuman world" and "seeks to ensure that it, too, may unfold in many diverse ways." In the English-language literature of environmental political thought, hundreds of books and articles discuss this distinction. They offer dozens of subtle definitional variations and develop innumerable arguments for the superiority of one perspective or the other.

The contrast with environmental discourse in France is stark. There, debate between nonanthropocentrists and anthropocentrists is peripheral at best. In fact, no French scholar makes this distinction central to an understanding of the varieties of *French* ecologism. Luc Ferry's *Le nouvel ordre écologique* (1992a) only seems to be an exception. Although Ferry's critique of environmental philosophy depends on the anthropocentric/nonanthropocentric distinction, most of his alarm is directed at English-speaking ecologists, including Christopher Stone, George Sessions, and Aldo Leopold. His attempts to read a couple of French ecologists in light of "Anglo-Saxon" ecologism goes seriously awry,[6] for France has been a seedbed for green theories that, in varying ways, elude the categories of English-speaking environmental thought. The question raised by studying French ecologism is not who has the advantage in debates between anthropocentric and non-anthropocentric ecologists. The question is whether that debate really has to be the leitmotif of ecologism at all.

In this book I argue that the absence of this debate in France has kept the discursive field open for different strategies of *noncentered* ecological argument. Rather than feel bound to situate their views in relation to some theory of the ultimate ground of environmental values, French green theorists tend to study how conceptions of nature and human identity intertwine. They elaborate green thought more often by *reciprocally problematizing* "nature" and "humanity" than by refining the distinction between them. In this sense French ecologists could be said to posit *divided natures*. They maintain that what "nature" is shifts in relation to epistemological, social, and political-ethical changes. Noncentered ecologists see "nature" as multiform and as inextricably confounded with humanity's projects and self-understandings. They are attentive to how the very meaning of being human is tied up with our constructions of "nature." For that reason, they believe, political ecology can pursue its tasks lucidly only by becoming aware of the processes linking "nature" and human identity. Noncentered green theorists forswear rhetoric that reifies nature and fashion a program whose content is as much "social" as "natural," all the while seeking to protect sources of experience that enrich human identity.

Typically, French theorists express their conception of political ecology as a form of renewed humanism. More particularly, I shall argue, they draw on traditions of *skeptical* humanism. Ecological humanism, therefore, is quite distinct from the epistemologically confident anthropocentric humanism that English-speaking ecologists eye with scorn—a tradition that exalts humanity and gives it unquestioned supremacy over nature. French ecologists draw on indigenous intellectual traditions associated with Montaigne, Pascal, and Rousseau. They use those traditions to question facile assumptions about human "nature" and thereby to tone down the hubris of Cartesian humanism. Simultaneously, skeptics challenge the adequacy of every apprehension of the "nature" of the external world. Humanism becomes ecological when it opens itself to reflecting on how nature and humanity are mutually defining.

Theory and Linguistic Communities

In recent years a few scholars have argued for the need to pay more attention to cultural distinctions in the way environmental issues are framed in different countries (Fischer and Hajer 1999; Macnaghten and Urry 1998; Guha and Martinez-Alier 1997). This book is a contribution to such a project, with a caveat: it cultivates an ear for particular accents in the works of ecological political theorists, more than in expressions of popular culture or in the attitudes of environmental activists.

Why highlight cultural particularity in green theory? On the face of it, the more conventional approach seems reasonable. Nature, after all, is nature. It seems to be of no consequence whether environmental damage occurs in New York, in Nantes, or in Nairobi. The considerations brought to bear in evaluating the damage should be everywhere the same.

But matters are not so simple. To live in a distinct linguistic community is to inhabit a "lifeworld" (defined by Jürgen Habermas as "a culturally transmitted and linguistically organized reservoir of meaning patterns"[7]). And the contents of those cultural reservoirs can differ significantly. It is not hard to see how this can happen. Theorists in English-speaking countries frequently read one another's books; they critique one another in environmental journals; they meet in conferences; they exchange academic positions. A glance at the bibliography of any of the surveys of ecological thinkers reveals that the works of Americans, Britons, Australians, and Canadians cross one another's borders with barely a nod from an intellectual customs inspector. Yet works of French ecologism somehow have gotten lost in transit. To a certain extent, the converse is also true in France, where, although books by Barry Commoner and James Lovelock can be found in translation, the whole literature of English-language environmental ethics remains the province of specialized scholars (Larrère 1997) and has little resonance among French green theorists more generally.

As a result of such differences in the diffusion of ideas, the conversations of entire linguistic communities take on distinctive characters. Over the years, the ease with which conversation passes between thinkers allows certain modes of argument, a range of terminology, a sense of exemplary problems, and unintended partialities to build up almost imperceptibly. Theorists take for granted areas of government activity (or inactivity) that would be controversial in other communities, and ecologists absorb attitudes toward "wilderness" or "pollution" that are common among their compatriots but unusual in other nations. Even those who disagree with the prevailing assumptions find it necessary to construct their arguments to fit the contours of the debate. As a result, their contrarian views can end up being formatted by the very ideas they reject.

In effect, the prevalence of certain concepts and modes of reasoning within a linguistic community creates a *rhetorical field*. A rhetorical field favors pushing inquiry into certain territories while leaving others relatively unexplored.

I do not use the contrasting expressions "English-speaking ecologism" and "French ecologism" merely to call attention to the national or cultural origins of different thinkers. Much more than that, I use them to capture the sense in which a shared language has become the basis for broadly shared assumptions and patterns of environmental discourse in two linguistic communities.

The promise of doing systematic, cross-community comparisons of theory lies in its potential to expose widely accepted assumptions and to allow them to be challenged. I take seriously the idea that cultures are incubators and preserves of difference. As I see it, the purpose of detecting difference is not to sanction relativism. Difference invites comparison and, potentially, correction. Since the conversation of each linguistic community is incomplete in relation to a wider universe of discourse, each community stands to improve its understanding of issues by deliberately contrasting their fundamental ideas.

Cross-cultural comparisons have their dangers, too. They involve broad generalizations that can deteriorate into stereotypes. The risks may seem especially high when an argument throws together ideas from many different countries. Some may wonder whether British ecologism is entirely of a piece with its American cousin. Some may also suspect that what I call "French ecologism" really describes green thought coming out of *most* of the countries in Europe, where the environment has been altered by steady human habitation for thousands of years.[8] Moreover, generalizing about an intellectual phenomenon such as ecologism, which everywhere divides into numerous schools of thought, can easily run roughshod over different thinkers' carefully drawn distinctions. I try to minimize these dangers in several ways.

First, I try to approach the works of many significant thinkers closely enough to give a sense of their argumentative texture. This book is intended more as a work of political and social theory than as a work of intellectual history.[9] In each chapter I identify a number of philosophically related French thinkers who have written extensively and perceptively on environmental issues. In many cases, I also try to locate their arguments in the context of their larger oeuvre. Then, to develop comparisons at an individualized level, I examine French ideas in relation to the ideas of particular English-speaking green thinkers. At this level, there can be no question of claiming that these thinkers are representative of anything other than their own thought. Still, I have chosen them from a wide range of tendencies—including deep ecologism, social ecologism, liberal environmentalism, bioregionalism, and ecosocialism—in order to strengthen my contention that the typical argumentative patterns of English-speaking ecologism show up in unexpected ways in different theories.

In view of the vast number of English-speaking ecologists, however, it is impossible to review even a substantial fraction of them in a comparative work of this sort. Thus, I supplement my individualized investigations with broader assessments drawn from the synoptic works of scholars of green thought—Andrew Dobson, Robyn Eckersley, Brian Baxter, Tim Hayward, and John Barry, among others. Those authors review hundreds of books and articles (with very few exceptions, written in English) and themselves sympathize with different tendencies within English-speaking ecologism. Thus, it seems likely that any themes that are common to all of them can be regarded as widespread features of English-speaking ecologism.

Third, I take care to qualify my points in ways that respect other views without compromising the generality of my argument. Throughout the book I will call attention to claims coming out of one linguistic community that are reminiscent of those found in the other. I do not want to claim that English-speaking ecologists *without exception* are committed to centered theorizing. Nor am I saying that "French" perspectives can *never* be found outside of France. My thesis pertains to patterns of argument in two linguistic communities taken as wholes. These are patterns that usually pass unperceived, precisely because speakers *use* discursive styles rather than treat them as objects of study. Even where an idea typical of one community finds its way into the discourse of another, it is often inflected in unfamiliar ways, subtly reshaped, or pressed into the service of arguments that alter its significance. It is precisely because we can learn to discern these shifts in meaning that cross-community comparisons are worthwhile.

Acknowledging difference, of course, is not to equated with perceiving truth. I do not contend that France's contributions to ecologism are superior to anything written in English, and I often criticize French ecologists. I do hope to demonstrate, however, that French ecologists furnish insights and conceptual materials for alternative perspectives that could help extricate green theorists from endless debates over the real "center" of nature's value. Why might it be important to do so? I would suggest two reasons, one philosophical, the other practical.

Debating Centers

Centered ecologisms have the philosophical disadvantage of minimizing the important truth that the meaning of "nature" is highly variable and value laden. For some writers, "nature" refers to the conditions for the physical and psychological health of human beings. Environmental politics may be about enhancing the quality of urban life by reducing air pollution and adding green spaces. For pastoralists, it is not the city but the relative calm of rural life—its seasonal regularity and its proximity to organic realities of life and death—that constitutes a "natural" setting for human flourishing. For other writers, the cultivated countryside is hardly nature at all. Real nature is wilderness. It is a world untrammeled by man, mysterious, sometimes threatening, exhilarating in its beauty and awe-inspiring in its spontaneous, life-perpetuating complexity. From the point of view of scientific systems ecologists, "nature" is a vast, evolving, nested set of mutually supporting homeostatic systems. Avoiding disruption of Earth's life-sustaining systems has more to do with controlling the emission of greenhouse gases or protecting the continental shelves than with controlling urban pollution, preserving the countryside, or setting aside wilderness areas.

Is only one of these natures real? Are the others secondary or epiphenomenal? Is pastoralism, for example, an understandable but ultimately retrograde expression of nostalgia—a holdover from a pre-industrial world, destined to die away as scientific systems ecology enables humanity to subject every part of the planet to rational control?

My position is that the array of natures expresses profound divisions in our apprehension and evaluation of reality. Each perspective is a different way of seeing what "nature" *is*. In addition, values insinuate themselves into each view. One nature seems to suppose that satisfying material interests is the pre-eminent need of an organic being. Another regards spiritual expression as more fundamental. Some natures presume that a life lived in accordance with truth requires devotion to scientific norms of objectivity. Others see truth in more poetic intuitions of wholeness and interconnectedness. Put this way, it also becomes evident that our understandings of nature correlate to equally divided views of our own subjectivity. Our views of nature imply answers—often contradictory answers—to questions about the very meaning of life.

To enter the centered environmental debate, one must pay the price of admission. The characteristic preoccupations of both sides divert us from seeing what is problematic in both nature and humanity. That is, by assuming that one notion of nature is fundamental and then focusing on whether that nature contains qualities sufficient to elicit respect by human beings, theorists tend to suspend inquiry into the identities of both parties to the relationship. Anthropocentrists and nonanthropocentrists assume that the division between humans and their nonhuman environment is ontologically fixed and can serve as the foundation of ecological reasoning. In the process, they discount the significance of natures that lie outside their field of theoretical vision.

Were theorists not wedded to the idea of ecological centering, they might feel less constrained to devise a theory of value that fixes the traits of genuine environmental concern a priori. What might most impress them is the *irreducible* diversity of "natures"—both nonhuman and human—implicit in environmental practice.

Noncentered ecologists are in a better position to see that the varieties of protests against environmental degradation derive their unity not from a theory of value but from the fact that they all see "nature" as a *problem*.

And that, in itself, is an epochal shift. Serge Moscovici, one of France's earliest and most perceptive green theorists, sets the tone for his own ecumenical ecologism with the following observation (1990: 7):

> . . . the great new concern of our epoch is the question of nature. It is a question that catches us out, both when we consider our given conditions of existence from the point of view of the species, and when we reflect that science and technology have transformed us into one material force among others. . . . In short, the state of nature is not now just an economy of things; it has become, at the same time, the work of human beings. The fact is that we are dealing with a new nature.

Moscovici holds that never before has "nature," in so many guises, been so *consistently* and so *self-consciously* a focus of *critical* engagement. A violated nature has become an essential factor in multiple expressions of a world gone wrong.

The existence of a "new nature" in this sense is sufficient to challenge age-old traditions of political theory, even if no single nature is granted onto-logical primacy. Traditionally, the concept of "nature" has been central to debates about what it means to have a well-ordered society. Ancients argued about the perfection of man's "natural" virtues in the ideal city; modern liberals asserted that just societies preserve "natural" rights. But those natures were fixed and supposedly knowable. Even when "nature" obviously changed—as in modernity's great shift away from teleological and toward causational understandings of "natural" processes—a new, *true* nature was summoned to supplant an old and inadequate one. Centered ecologisms, I shall argue, for all that they challenge earlier understandings of nature's ability to absorb human-induced changes, continue this tradition.

Noncentered ecological theories of the sort commonly encountered in France, on the other hand, problematize the very founding concepts out of which environmental concern emerges. For Denis Duclos (1996: 301), "ecology is at the heart of today's philosophical, anthropological and political problems because it sends us back . . . to the question of the limits of human practices." Crucially, Duclos contends not that ecologism defends a determinate "nature" but that its various claims all raise the idea that there are goods we can never secure by continually extending our control over our surroundings. "Nature" has become the vehicle for expressing a vast array of worries about the quality of life. Environmental philosophies shortchange this truth when they force us to choose one nature rather than another.

Centered theories also have a practical, political drawback. Polarizing humanity and nature, they create obstacles to imagining a political program that combines typically "environmental" concerns (e.g., preserving species and rainforest ecosystems) and sensitivity to broader issues such as social justice and the meaningfulness of everyday life (e.g., redistributing work, democratizing environmental risk decisions, improving life expectancy in developing countries).

Robert Goodin (1992) clarifies how tensions can arise between policies that often coexist in green parties' programs. Goodin advocates one version of nonanthropocentrism. He is fully aware that the programs of the green movement have included seemingly human-centered demands to make society more democratic, egalitarian, pacifistic, and multicultural. Seeing no way to derive such ideals from his theory of nature's intrinsic value, he maintains that their appropriateness as components of a green platform must depend, contingently, on whether or not they contribute to "producing good green consequences" (ibid.: 16). Asserting "nature's intrinsic value" thus puts various human values in a strictly subordinate position. Goodin draws the implication of that subordination: if participatory democracy or a commitment to nonviolence turns out to impede the achievement of green goals, it is the former that must give way, not the latter (ibid.: 120). Nonanthropocentrism can validate principled refusals to compromise nature's value.

Anthropocentric environmentalists tend to be more pragmatic. They are often more willing to seek a balance among goods, environmental or otherwise, insofar as all goods are weighed on the scales of human interest. Considerations of democratic participation, social inclusiveness, and cultural diversity mix with environmental concern. Each consideration is given its due out of respect for justice. Nonanthropocentrists remain dissatisfied. They suspect that people keep slipping their opposable thumbs on the scales of justice. They worry that the very process of balancing values leaves environmental goods too vulnerable to continued exploitation.

Such theoretical divisions lead all too easily to bitter rivalries among those who might otherwise have substantial grounds for political cooperation. So severe have strains within centered ecological theory become that they motivated Brian Norton to dedicate a whole book to working "toward unity among environmentalists" (1991). Significantly, however, Norton cannot end his book without revisiting the philosophical question that causes such disunity. At that point (p. 255), he decides for anthropocentrism! Thus, it may be that the only way to avoid lapsing into centered ecological debate is to take a different philosophical path from the start. That is what French ecologists do.

The prevalence of noncentered ecologisms in France stands to alter our understanding of ecological discourse in three ways.

First, within the writings of a single theorist, claims get developed independent of their potential contribution to either nonanthropocentrism or anthropocentrism. This is not to deny that readers will encounter green claims that are familiar to them from the literature of English-speaking ecologism (e.g., demands for holistic thinking and environmental justice and for caution in the application of technology). The difference arises in the way such claims get framed theoretically. As I shall show at many junctures, English-speaking ecologists tend to press the most varied observations—about ecological systems, about animal behavior, about the social construction of nature, about environmental justice—into the service of one value center or another. The French help us to see how familiar ideas can lead in new directions when they are no longer under this rhetorical pressure.

Second, exchanges between two or more French theorists suggest how ecological debate proceeds when the philosophical reach and consistency of a theory of environmental value are not the main issues. Noncentered thinkers challenge one another to confront the significance of their conceptions of nature and humanity for the distribution of power in a community. Every conception of nature, it turns out, has implications for how control is exercised over nonhumans and humans alike. Noncentered theorists are especially adept at teasing out such implications and subjecting them to critical scrutiny.

Third, with regard to the whole field of green political theory, adding French thinkers into the mix may rebalance our perception of its dominant controversies. Their presence may help dispel Goodin's (1992: 8) impression that "the insight that drives, most powerfully, the current wave of environmental concern" is that nature has "an independent role in the creation of value." A proper acknowledgement of French ecologism might help reorient green political theory generally away from interminable debates in environmental ethics. It might lend credence to a claim only occasionally heard and even less often heeded in the English-speaking world: that the future of green theory lies beyond anthropocentrism and nonanthropocentrism.

The Varieties of French Ecologism

French ecologism, like its English-speaking counterpart, is far from homogeneous. Green political thought splits into a number of variants that share philosophical kinship. Although no two taxonomies are identical, English-speaking scholars typically break green thinkers down into categories such as deep ecology, social ecology, ecosocialism, ecofeminism, postmodern ecology, and bioregionalism. Only one of these categories—ecosocialism—fits the French case neatly. French ecologism is best understood in terms of different ways of thinking about divided natures.

Each distinctive strain of French ecologism consists of a number of thinkers who share two things. First, by virtue of agreeing on certain methodological and ethical premises, they offer a common interpretation (broadly speaking) of the reciprocal implication of humanity and nature. One group sees it as a historical process of cumulative technological and social transformations; another points to conceptual crossovers between scientific and humanistic understandings of nature; another explores the psychological impulses driving different ways of constructing the nature/culture divide. Second, the distinct strains correlate to shared visions of the political implications of ecologism. Personalists, for example, believe that ecological ends are best served when political power is widely dispersed. Politicizers, in contrast, imply that overcoming environmental degradation requires gathering representatives of humans and nonhumans together in a more centralized legislative assembly. The diverse strains of French ecologism testify to a range of conceptual possibilities that open up when theorists ponder environmental challenges independent of a commitment to the center of value.

The French owe their independence to the peculiar intellectual and social circumstances of the birth of political ecology on their soil. Chapter 1 traces the origins of a rhetorical field in which noncentered ecologisms would develop. In a country without a strong tradition of protecting wilderness, environmental concern developed relatively late and in the context of an extremely wide-ranging movement of political contestation. As a result, the French ecology *movement* developed a program that addresses "social" issues as much as "environmental" ones. The early leaders of the movement asserted a connection between the "natural" and the "social"; however, they sometimes left the impression that they saw a contradiction. French green *theorists* have taken it as their task to devise a "humanistic" ecologism that avoids this potential dualism.

Each subsequent chapter pits a philosophically related group of French thinkers against certain English-speaking counterparts to show how reconceiving the *connection* between humanity and the world promotes an ecologism that stays attentive to divided natures.

This sorting of thinkers into varieties begins in chapter 2, which investigates the nature of "nature" as developed in the work of Serge Moscovici, France's most prescient green thinker. Moscovici's work exemplifies an ecologism in which "nature" and "society" get constituted historically out of human interaction with the material world. I distinguish such an orientation from one that defends nature's intrinsic value. Proponents of nature's intrinsic value are particularly hard pressed to explain how human beings can avoid imposing human interests on "nature" even as they seek to protect it. Moscovici's viewpoint differs from anthropocentric ecologism, too, because of his insistence that human subjectivity and nature are interdependent. Noncentered ecologists, I argue, displace the puzzles of centered theory by skeptically questioning both the unity of human reason and the knowability of the external world.

Chapter 3 takes up the pervasive influence of systems theory in French ecologism. Systems theory is a holistic approach to the study of goal-oriented entities: cybernetic machines, living organisms, entire ecosystems. Systems ecology helps anthropocentrists understand how far humanity can go in using ecosystems before thresholds are crossed that send them into decline. Nonanthropocentrists sometimes argue that systems theory does far more than identify nature's limits: that it grounds a new ontology of nature's intrinsic value. French theorists, in contrast with centered theorists, use systems theory more to unify understanding in the natural and social sciences and, at their most ambitious, to express a new theory of the historical development of ecological rationality. These efforts culminate in the monumental oeuvre of Edgar Morin, who tempers other French thinkers' enthusiasm for systems theory with the realization that even the most sophisticated cybernetic models of ecosystems belong in an open-ended series of ways of knowing nature.

"Politicizing" theorists, including Michel Serres and Bruno Latour, are the topic of chapter 4. Such theorists contend that political concepts (e.g., law, power, hierarchy) run through our conceptions of nature and technological risk. For these thinkers, humanistic disciplines such as literary studies, philology, and philosophy shed light on the representations of nature that structure environmental thought. Theorists of this type end up calling for more deliberative ways of setting up interactions between human communities and their environments. At times this approach appears "postmodern," but postmoderns who assert the utter incommensurability of values tend to relativize the very scientific knowledge that sparks much environmental concern. Latour and Serres, in contrast, exemplify an ecologism that questions science skeptically but steers shy of depicting it as only a form of knowledge/power.

Chapter 5 takes up applications of "personalism" to political ecology. Personalism is a spiritually oriented philosophy that was most powerful in the 1930s, when it was espoused by Emmanuel Mounier. Mounier's critique of modernity emphasized the dehumanizing effects of advanced technologies and the homogenizing consequences of bureaucratic social organization. The philosophical contribution of personalist ecologism is to suggest that the multiple objects of environmental concern designated by the term "nature" (e.g., pristine landscapes, healthful consumer products, complex ensembles of spontaneously evolving phenomena) correlate to different aspirations of the human personality. Denis Duclos's studies of how "nature" appears in relation to the passionate, decentered individual offer the most compelling updating of secularized personalist insights.

"Ecosocialism," the topic of chapter 6, is a strain of environmental theory common to France and the English-speaking world. Ecosocialists protest the relations among resource depletion, alienating work conditions, and the unjust treatment of Third World countries. Some ecosocialists have created unresolved tensions between centralizing and decentralizing approaches to ecological reform; others have championed relativistic theories that fail to translate the moral urgency of environmental concern. In recent years, however, a new strain of ecosocial discourse has taken shape in France. Social activists such as Jean-Paul Deléage and Alain Lipietz favor a contractual ecosocialism in which ideals of equality and autonomy are conceived as the fundamental values of ecological negotiators who seek to win the assent of diverse groups to a social order that is stable, distributively just, and environmentally responsible. English-speaking ecosocialists typically get drawn into debates over nonanthropocentrism. French contractual ecosocialists are freer to explore how human appreciation of nature is mediated by historically evolving modes of labor.

Chapter 7 situates ecologism in relation to liberal thought. Liberals sometimes worry that an ecologistic worldview is inherently undemocratic. They charge that ecologists assume that what is natural is good, thereby denying human communities the right to set their own purposes. Liberals contend that the distinction between nature and culture must be preserved if human freedom is not to be endangered. Too often, however, French liberals and French ecologists have allowed their debate to be stalled by hyperbolic mischaracterizations. The liberals have ignored efforts to devise a noncentered ecologism; the ecologists have not always faced up to the liberals' contention that humane and democratic theorizing cannot avoid centering on human reason. In chapter 8 I draw together the arguments for a more conclusive debate. Even in noncentered ecologisms, I maintain, it is possible to detect traces of some sort of rationality that contradicts the language of contingency and arbitrariness preferred by certain French theorists. I show how Habermas's theory of discursive ethics might account for those traces without requiring ecologists to abandon their insights about the reciprocal implications of "humanity" and "nature." At the same time, I argue, France's skeptical humanist heritage supplements the theory of communicative competence. It offers up an ideal of the political ecologist as a crossbreed whose ability to move among the worlds of scientific, humanistic, and social inquiry helps keep rationality balanced by insisting on the reality of divided natures.

1
Problematizing Nature

"In many respects," writes Jean-Luc Parodi (1979: 40–41), "[French] ecologism appears, when it finally became structured, as the ideological heritage of May 1968." According to Parodi (ibid.), May 1968 remodeled France's ideological landscape by "favoring themes of worker self-management, spontaneity, the criticism of bureaucratic apparatuses, by establishing 'secondary fronts' in feminism, regionalism." In contrast to countries that have preservationist and conservationist traditions stretching back into the nineteenth century, France has not typically treated preserving nature per se as a public concern.

The simultaneous appearance of new social and environmental demands in the late 1960s profoundly affected how French ecologists problematized nature: they now drew inspiration from a sweepingly critical discourse about society as a whole. French ecologism developed in a process that Moscovici (1993) calls a "polymerization" of ideas. Five strands, each a part of a "culture of life," intertwined to oppose the "forces of destruction" in the contemporary world. One strand was a critique of science, particularly of its development of atomic weapons and its integration into economic and military structures. A second drew from a critique of colonialism and the destruction of ethnic diversity worldwide. A third, in the name of "human authenticity," updated a European sensibility favoring the countryside over the city. Moscovici calls the fourth strand sociological, since it links ecology to class structure and work relations. Through it, Marxist analyses of relationships between science and industry or of the state's preference for economic growth over social justice became a standard part of French ecological discourse. Finally, a "mixed network" of biologists and cyberneticians began to challenge physics-based understandings of the natural world, replacing mechanistic models of relations between things with models that emphasized mutual interdependence and self-regulatory processes.

Notably absent from this list, especially from a New World point of view, is any notion of saving what is wild—of protecting "nature" in the form of territories, landscapes, and species unaltered by human activity. In English-speaking ecologism, a devotion to the "wild" often shows through at crucial points and gets formalized in a nonanthropocentric environmental ethic. Anthropocentric ecologists may criticize the coherence of nonanthropocentrism and seek to combine environmental concern with social theory. Yet in English-speaking ecologism efforts to defeat nonanthropocentrism have an unintended effect. They accentuate what anthropocentrism and nonanthropocentrism have in common: a preference for *centered* environmental theory. Centered theories ground environmental values in an entity—the human subject or nature—with a distinct and stable identity.

The polymerized French critique, in contrast, prefigures an ecologism in which "nature" and "humanity" constantly criss-cross, making their identities plural and unstable. Not unlike greens elsewhere, popularizers of French ecologism have long assumed that political ecology is somehow of a piece with advancing greater social equality, workplace democracy, participatory politics, and self-realization. Targeting "productivism" and espousing "autonomy," they developed a discourse that consistently mixed ecological and social concerns. But unlike their counterparts in the English-speaking world, French ecologists have not set about perfecting a specialized ethical discourse to establish the priorities of an authentic green politics. As a result, French ecologism has never taken the form of a philosophical showdown between two "centers."

The belated attention to environmental matters in France has meant that French ecologists have had to make strenuous efforts to justify treating ecology politically. A reticence about linking ecology and politics has characterized not only relatively conservative nature-protection groups but also (for different reasons) post-1968 left libertarians. Even the enthusiasts of France's green movement have often been reluctant about organizing its adherents into durable structures capable of making binding collective decisions and then engaging in partisan electoral activity. René Dumont became a model figure for French ecologism because of his determination to surmount these tendencies. Through multiple roles (candidate, counselor, gadfly, publicist, protester, political intermediary), Dumont created such a striking example of the *practice* of political ecology that no study of what it means for ecology to be political can afford to ignore him.

Unusual by the standards of a country where intellectuals have often influenced the priorities of radical politics, the French ecology movement developed, for the most part, independent of the influence of theoretical systematizers (Faucher 1999: 38–39; Villalba 1997: 85–89). This innocence of theory has fostered ideological ambiguity. The popular representatives of political ecology simultaneously problematize nature and humanity, but they do so in ways that make the theoretical unity of their program questionable. This helps to explain why French ecological theorists often criticize the green movement. Yet by problematizing nature in the context of an effusion of social ideals the green movement maintained the baseline for theorists bent on grounding a new ecological politics in the *juncture* of humanity and nature.

Beginnings

In 1872, in an event that is often seen as a turning point in the history of political ecology, the United States created Yellowstone National Park. According to Roderick Nash (1967: 108), this was "the world's first instance of large-scale wilderness preservation in the public interest." The authority of the federal government would protect Yellowstone's natural wonders against private appropriation and possible destruction. This was the first time that authorities acting in the name of the whole people saw fit to intervene against private interests to protect things—geysers and spectacular vistas—not created by man and having no immediate utility. Moreover, protection was justified precisely *because* of those qualities.

That such action became desirable owed much to Henry David Thoreau and to John Muir (Taylor 1992; Nash 1967: 84–107). Thoreau's conviction that "in Wildness is the preservation of the world" and Muir's rapturous vision of nature as sublime beauty appealed to a nation that had learned to associate frontiers with freedom and unspoiled landscapes with God's love. For Muir, our kinship with nature as a whole made the wilderness a source of spiritual renewal. Preservationists popularized the idea that natural things are valuable for what they are. The preservation of the sublime beauty of wilderness was therefore justified for reasons that went beyond mankind's need for resources or recreation and beyond a concern for human health. Around 1900, the idea that government intervention to protect exemplary fragments of original nature was legitimate and perhaps even obligatory began to capture imaginations far beyond the United States. Yellowstone already had become a model for preservationist initiatives in Canada, Australia, and New Zealand.

English-speaking authors' histories of political ecology typically document that preservationism has competed with "conservationism" to justify moderating human uses of nature (Norton 1991: 6–13; Taylor 1992: 1–26; Eckersley 1992: 33–47; Paelhke 1989: 14–22). Gifford Pinchot, appointed chief of the U.S. Forest Commission in 1896, worried that the United States was wastefully depleting its resources. Forests felled without replanting, for example, would deprive the nation of the wood it needed for future development. Resources were important for their utility to human beings; therefore, they should be managed rationally on the basis of scientific studies and according to principles of economic efficiency. It was perfectly consistent with this view to see the forest as "strictly . . . a factory of wood" (Pinchot, quoted on p. 19 of Taylor 1992). Wisely managed, resources would ultimately contribute to raising the material standard of living of all citizens.

As preservationist and conservationist rationales for environmental concern were gathering force in the New World, the French were still preoccupied with fundamental issues of constitutional consolidation. Not the question of the state's competence to safeguard resources or to preserve wilderness but its very legitimacy was yet to be resolved. Viard (1990: 93) vividly captures the contrast:

When Americans were inventing the first national park, Paris had just lived through the Commune, the villages of southern France had reached their maximum population. . . . The Empire [of Napoleon III] had fallen, France was occupied [by Bismarck's Germany]. Americans were discovering the West and occupying it physically and symbolically. In France, people were endlessly debating and fighting for or against the Republic. . . . France was coming to terms with the destruction of the Bastille when Americans were confronting nondomesticated nature.

In France, the idea that it might be a public responsibility to protect parts of the nonhuman world would barely take hold before the middle of the twentieth century.

In the nineteenth century, France's difficulties in "coming to terms with the fall of the Bastille" left "nature protection" to be dealt with as a private matter. Professional associations (especially scientific ones) and interest groups occasionally acquired and managed relatively undeveloped sites. The project of "protecting nature" in France is usually traced to the founding of the Société impériale zoologique d'acclimatation in 1854 by the naturalist Isidore Geoffrey Saint-Hilaire. This society and other associations of botanists and zoologists sought to protect plant species and birds from the devastating effects of industrialization on their habitats (Simmonet 1982: 100). Usually their members' motivations were more scientific than aesthetic or utilitarian: relatively undisturbed habitats should be preserved so that scientists could accurately study ecosystemic interdependencies. In addition, in the middle of the nineteenth century, under the impact of urban industrialization, the "cultivated classes" increasingly sought relief in secondary residences in the country. There they acquired an interest in bucolic sites for recreation and relaxation (Fabiani 1986: 33).

Beginning in the late nineteenth century, these interests led various organizations to establish nature preserves. Artists of the Barbizon School of countryside painting saw to it that part of the Forest of Fontainebleau became an "aesthetic reserve" (Stevens 1991: 26). In 1912, Brittany's Ligue Française pour la protection des oiseaux established a preserve in the seven islands of the Côtes du Nord (Claude Lachaux, cited on p. 91 of Viard 1990). A larger preserve in a salt-marsh area called the Camargue was created in 1927 by an agreement between the Société nationale de protection de la nature and a salt-processing company (Davis 1988: 9). Such groups kept their distance from France's centralized political processes and did not advance a vision of the social order as a whole (Journès 1979: 233). They emphatically refused to support particular candidates (Abèles 1993: 8).

The debating and fighting of nineteenth-century French politics continued well into the next century. Nothing like Theodore Roosevelt's environmentally progressive policies took shape in France, which faced World War I, social unrest in the 1930s, defeat at the hand of the Nazis, the shame of the collaborationist Vichy regime, and the political instability of the Fourth Republic. Not until 1957 was a law passed placing restrictions on the development of property within areas designated as nature preserves. The Fifth Republic, declared in 1958, brought the first regime that could systematically address new public sensibilities in the area of environmental protection. France established its first national park in 1963. The first comprehensive national legislation to protect water basins took effect in 1964.

On France's anciently civilized territory, nature protection could barely be imagined as a matter of separating some sort of pure nature from human influence. Debates leading up to the adoption of the law allowing the creation of national parks emphasized that the French conception of nature differed from the conceptions of nature prevalent in other (especially Anglo-Saxon) countries. One member of parliament observed that French national parks cannot protect "virgin" nature, since "it is impossible to find a single significant, extended area in France" that remains innocent of human intervention (Viard 1990: 102). An official explained that national parks *à la française* would represent what nature *should* be (ibid.: 103). Nature protection would have to be conceived not as a matter of setting aside wilderness but as a new way of weaving together human design and biophysical processes. This new way allowed scientific, recreational, aesthetic, and historic preservationist reasons to count in defining the public interest in areas where private property had reigned supreme. Still, legislative debates of the time gave no indication that environmental concerns might radically challenge conventional notions of economic development or inspire a new political movement.

Only a few intellectuals foresaw the problems to come. One of the earliest expressions of environmental reflection in France (at least, one of the earliest of the expressions that continue to be sources of inspiration in environmental circles[1]) came in 1937 from a militant in Emmanuel Mounier's personalist movement. In his essay "Le Sentiment de la nature," Bernard Charbonneau argued that mankind should turn away from battling against nature and devote itself instead to struggling against "social determinisms" (Cans 1992: 203). According to Charbonneau, the means of social control had advanced to the point where they—not the forces of nature—were the main impediment to human liberty. In fact, Charbonneau claimed, contact with nature was the best stimulant to free individuality.

Heightening a "feeling of nature" was the aim also of a Swiss painter and intellectual who has had some influence in France's environmentalist milieux. Robert Hainard, in his books *Et la nature? Réflexions d'un peintre* (1943) and *Nature et mécanisme* (1946), provided the terms for a wilderness-loving—and very conservative—version of French ecologism.[2] He championed artistic forms that would stimulate a nonrational, sensual appreciation of nature's mystery and wholeness. He criticized humanism for aspiring to free humanity from all constraints. Instead, he declared, freedom should be conceived as "obeying strong and strict determinisms." In spite of such ideas, Hainard did not really push French thinking far in the nature-centered direction that his words sometimes implied. His seeming anti-anthropocentrism broke down in an admission that he wanted to "protect nature not for itself, but for the joy it gives man." Jacob (1999: 89, 100, 108) notes that Hainard sometimes made nature and culture opposites, sometimes saw them as complementary, and sometimes depicted human culture as within nature. Such philosophical confusions have attracted no major figure of French ecological theory to dwell on Hainard's ideas. Nonetheless, Hainard deserves a place in the history of French ecologism for having been one of its few thinkers to articulate the nature-protection societies' desire to create natural parks. Defining nature as the world acting by itself, outside of human intervention, Hainard declared that "nature, wild and free, will be the engine of social reform." What that might mean in practice was foreshadowed in the career of another precocious naturalist: Roger Heim, author of *Destruction and Protection of Nature* (1952). In 1957, outraged at the waste of natural resources, Heim took the lead in protesting plans to build a highway through the Forest of Fontainebleau (Acot 1988: 228; Jacob 1999: 171).

Also in 1957, the expression "écologie politique" was coined by Bertrand de Jouvenel (Duverger 1992). In the midst of France's postwar economic boom, Jouvenel scandalized economists by questioning whether societies governed by "an ideology of growth" had a wide enough conception of the quality of life. A "productivist" society, he said, is organized to produce an "abundance of temporal goods." To do so, Jouvenel argued, it encourages the mobility of labor and the formation of "opportunist consumers"—economic agents whose desires revolve around low-priced consumption goods. One additional effect of a productivist orientation is that its system of accounting fails to register damages to the natural world unless they directly affect people's property. Society thus becomes a "parasite of its environment." A more complete and defensible economics would have to acknowledge that "production has two forms, one with a positive value, the other with a negative value." An avid reader of Rousseau, Jouvenel suggested that a wider conception of well-being might contain such goods as community attachment, spiritual commitment, and personal moderation. Fostering such well-being would require passage from "political economy" to "political ecology."[3] Within a few years, perceptions of a material crisis would add urgency these early reflections.[4]

The work that is most often seen as responsible for first popularizing political ecology in France was published in 1965 by Jean Dorst, a professor at the Museum national d'histoire naturelle in Paris (Vadrot 1978: 26; Acot 1988: 229). *Avant que la nature meure* [*Before Nature Dies*] is a veritable catalogue of worldwide environmental ills—and, in response to them, a call for a "political ecology." Dorst decries the human population explosion, soil erosion, the destruction of aquatic habitats, the spread of insecticide residues, and oil pollution in the seas. He explicitly rejects the idea that creating private "nature preserves" will suffice to preserve nature: "If one wishes to save wild nature—or at least what remains of it—this can only be done by integrating it in the theater of human activities. Because of the unity of the world, any solution must be applied to the totality of the planet, whose development man must foresee as a function of his interest rightly conceived." (Dorst 1970: 16) Although the reference to man's "interest rightly conceived" seems to put Dorst firmly in the anthropocentric camp, elsewhere he suggests that protecting nature is more than a matter of human interest. Man, he declares, has "no moral right" to extinguish a plant or animal species. Certain ethical duties to nonhuman species circumscribe the pursuit of human interests.

Dorst imagines that solutions to the problems he reviews will result from a "realistic entente between economists and biologists." These experts will ensure "the rational development of humanity within the framework of harmony with natural laws" (ibid.: 18). The difficult balancing of values involved in ecological politics would thus take place more technocratically than democratically. Questions about the shortcomings and dangers of technocratic environmentalism receive no attention from Dorst. Would citizens resent having their lives "managed" by technocrats? Might biologists' views of "natural laws" be skewed by questionable social prejudices? Is it possible that economists' views of "rationality" rest on one-sided views of human nature?

1968: Political Uncertainties of a Nascent Ecology Movement

The next stage in the development of French ecologism arose out of a rebellion against every form of technocratic thinking. In May 1968, after 10 years of paternalistic government dominated by General Charles de Gaulle, French university students and then workers took to the streets, occupying factories and parts of central Paris. Their target, recalls Alain Lipietz (1989: 68), was "an omnipresent, technocratic state." With future German Green Daniel Cohn-Bendit as their charismatic spokesman, students protested against the centralization, authoritarianism, and exclusivity of the university system. Workers demanded a larger share of their society's burgeoning wealth and more control over their working conditions. Guy Debord, Raoul Vaneigem, and other "Situationists" of 1968 criticized urban gigantism, loneliness, and the pursuit of quantity of goods rather than quality of relationships in everyday life (Simmonet 1982: 54, 94). They railed against routinization, hierarchy, and citizen passivity. Direct political engagement—outside the rigid structures of the established parties—was the order of the day.

Ecological themes entered the ideological mix gradually (Legoff 1981). Dorst's *Avant que la nature meure* and Rachel Carson's *Silent Spring* (published in French in 1963) took on new significance after 1968. They served as elements of a general critique of a society so hedonistically dedicated to material consumption that it was willing to wallow in its own wastes and to tolerate the commercialization of spectacular landscapes. The *Torrey Canyon* oil tanker disaster of 1967, which fouled the coasts of France and Great Britain, had already made the environmental dangers of modern technology more vivid to the French than ever before. France's first national park had just opened (in the Vanoise, in 1969) when plans were announced to allow a ski area to be built there. The journalist Jean Carlier led a petition drive that convinced President Georges Pompidou to stop it.

Not that Pompidou had himself become a convert to a new ecological sensibility. He introduced plans to build a highway on the left bank of the Seine river in Paris with the admonition that "the city must adapt to the automobile"(Pronier and le Seigneur 1992: 141). Indignation at that project occasioned one militant's turn toward ecology. Brice Lalonde—president of the student union of the Sorbonne in 1968, and the man who would become political ecology's most electorally ambitious advocate in the coming decades—organized a massive bicycle-based demonstration in Paris in 1972. In the early 1970s, the growth of a feisty ecological press, with magazines such as *Charlie Hebdo, Combat Nonviolent, La Gueule Ouverte*, and *Le Sauvage*, began to amplify more militant voices (Eisendrath 1979: 215–216). Such publications, more than structured organizations, were instrumental in bringing a wide range of activists together in France's two most significant environmental protest movements of the early 1970s: the opposition (beginning in 1970) to the construction of a nuclear power plant at Fessenheim and the 1973 confrontation over the extension of a military camp at Larzac (Bennahmias and Roche 1992: 23–34).

Pierre Fournier, an ecologist who wrote for *Charlie Hebdo* and *La Gueule Ouverte*, was among the first to call for a demonstration at Fessenheim. He argued expansively for a "a new ecological left" that would be "the sole genuine extension of the great, liberating burst of laughter of May 1968" (Allan Michaud 1989: 105). Fournier (cited on p. 62 of Sainteny 1991) spoke of a revolution that would move France away from an "economy of growth to an economy of equilibrium and sharing." His writings inspired some to "return to the earth" in rural communes, to create autarchic societies practicing artisanal production and "organic farming" (Simmonet 1982: 90–91, 95). His libertarian inclinations led him to ground an ecological sensibility in an appreciation of the potentialities of life—especially one's own life. Changing society was, first of all, a matter of liberating one's own passions.

The anti-nuclear movement of the 1970s is what first brought together activists of the New Left, libertarians, and naturalists. Nuclear power symbolized an extraordinary range of evils opposed by the new social movements: dangers of accidents and pollution, industrial gigantism, centralized production, extreme artificialization and control as opposed to spontaneity and respect for nature. Protests against nuclear power seemed to embody a new egalitarianism, too, insofar as they attracted participants from across social classes and partisan divisions (Anger 1981: 35–36). Combining all of these motives and values, the anti-nuclear movement helped politicize ecology. If in the past it had been possible for the nature-protection societies to safeguard their nonpartisan image by avoiding most confrontations with the state, no such luxury was available to those who challenged nuclear power. To question nuclear power was inevitably to challenge the state's authority in setting the nation's priorities, and the French state was determined to press ahead with its nuclear power program, protests or not. Economically, nuclear power represented a virtually inexhaustible means of powering economic development; militarily, it secured France's independent nuclear defense strategy.

At the same time, the anti-nuclear movement's ambiguities made it an easy target for critics (Faivret et al. 1980). First, the movement was not clear about what, exactly, it was opposing. Was it a particular power plant? Or was it France's entire nuclear power program? Was it a particularly dangerous technology or amoral scientism more generally? Was it the power of a distant, technocratic state that could impose its development without democratic consultation? Was it an entire strategy of economic development, of which nuclear power was only one instance? (Vadrot 1978: 60–61) A sociological study of the anti-nuclear militants by Alain Touraine found proponents of all of these positions and many others. With obvious disappointment, Touraine (1983: 195) concluded that the anti-nuclear movement was "precariously balanced between positions of principle and concrete but limited actions, a situation which prevents it from playing a specifically political role." The movement's second weakness concerned power and organization. The libertarian sentiments of 1968 created an abiding aversion among activists to legitimate permanent structures—even if those structures would be necessary to develop a disciplined, coherent, comprehensive plan to carry out social transformation. Those inspired by the social movements of May 1968 certainly had no hesitation about launching broad social critiques and confronting the state's power; however, they rejected much of what is ordinarily meant by "political action." Insistence on personal autonomy and decentralization worked against the construction of anything resembling a permanent party (Sainteny 1991: 18; Pronier and le Seigneur 1992: 50–51). Ecological activists chafed at the very ideas of authority and organizing power. Brice Lalonde, who later ended up doggedly pursuing an electoral environmentalism, asserted in the 1970s that "a party is not really an ecological institution" (Parodi 1979: 38). Parties signified hierarchy, not equality; compromise, not principle; programs and discipline, not critical freedom and spontaneous expression. True politics, in this view, consists in diverse forms of nonviolent popular mobilization, without durable or centralizing organizational structures, motivated by a vision of a public good that is promoted consensually. From the theatrical student protests of 1968 to the massive anti-nuclear demonstrations at Fessenheim, Flamanville, and Malville to instances of civil disobedience and ecological gendarmerie in the 1980s (Whiteside 1992: 19), French greens have long claimed to be engaged in a different way of doing politics.

Can ecologism be *effective* without partisan engagement? Ecologism is not only a movement dedicated to improving the condition of its adherents' souls. It loses its point if it cannot transform human activities so as to improve environmental conditions. Certainly the demonstrations in the 1970s drew tens of thousands of anti-nuclear protesters (Bennahmias and Roche 1992: 43–50), but ultimately these loosely coordinated efforts turned out to be no match for a state determined to press ahead with its goal of gaining energy independence from the unstable Middle East. Many in the French environmental movement refused to see that moving toward an alternative society might take years—indeed decades—of tiresome organizing and persuading, including electoral struggle, not just sporadic outbursts of critical energy. There might be additional challenges to making ecology political.

The Project of Politicizing Ecology

The first stirrings of a political organization for environmentalists occurred around 1971, when Alain Hervé formed a French branch of Friends of the Earth; Brice Lalonde would soon become its leader (Pronier and le Seigneur 1992: 141–143). In Alsace in 1973, a young biologist named Antoine Waechter helped found France's earliest version of an ecology party: Écologie et Survie (Buchmann 1990: 7). Beginning in 1978, Philippe Lebreton[5] was instrumental in organizing support for ecology candidates in various national elections (Jacob 1999: 140–142). But the idea of making ecology a committed political force gained ground largely due to the efforts of another individual.

In 1973, a number of journalists and certain members of Friends of the Earth began to talk of proposing an ecology candidate for the upcoming presidential elections. They sought someone already known by the public, someone deeply concerned with ecological issues, and someone with experience in the rough and tumble world of political engagement. From the start, their virtually unanimous choice was René Dumont—a former Trotskyite, a world-famous agronomist, and an activist on behalf of the Third World. Dumont treated ecology's 1974 foray into the electoral arena with almost deceptive matter-of-factness. He explained: "I believe that we must use all means of expression that are given to us to affect public opinion: there must be no limits to communication! Also it seems normal to me to intervene in the electoral process. . . ." (Dumont 1978: 185). From his own experience, though, he knew that it was far from normal.

Not even people with environmentalist sympathies gave much support to Dumont's candidacy. Remaining true to their traditions, the Fédération française des sociétés de protection de la nature and the Ligue pour la protection des oiseaux refused to join his cause. They judged his campaign too politicized (Vadrot 1978: 46). Many on the left opposed Dumont's candidacy because it threatened to drain away votes just when, for the first time in the Fifth Republic, the Socialists and the Communists had agreed on a Common Program that might enhance their electoral fortunes. Dumont received only 1.32 percent of the first-round vote in the election. After his quixotic campaign, the rest of the 1970s were years of "difficult consolidation" for the French green movement (Sainteny 1991: 13–22). Committees and networks set up to support Dumont disbanded or lapsed into inactivity until the next election. Environmentalists did have a few electoral successes in the late 1970s. Solange Fernex, Didier Anger, and other anti-nuclear activists drew ecology's highest electoral scores to date in municipal elections in 1977 and in legislative elections in 1978.[6] But until 1984, failure dogged every attempt to form a national green party with permanent statutes, elected officers, and an authoritative electoral program. Nonetheless, Dumont's candidacy marks a turning point in French ecologism.

Deliberately challenging his fellow citizens' long-standing aversion to partisan environmental activism, Dumont made it his personal, decades-long crusade "to ecologize politics and to politicize ecology" (Dumont 1978: 185).[7] That project entailed far more than running an electoral campaign with environmentalist themes. It meant casting ecological problems in such a way that a public inclined to deny their seriousness would finally give them political standing. It implied, additionally, bringing fellow ecologists to see their favored issues not in isolation but contextually, in terms of the flow of power in the community as a whole. Finally, since communities expect power to be exercised rightly, it affirmed that ecological issues could not be separated from issues of social justice. After the 1974 presidential campaign Dumont would never again serve as the leading candidate of the ecology movement (Bennahmias and Roche 1992: 42), but over more than 25 years he instilled a respect for politics in the French green movement.

Dumont's first politicizing lesson was that appreciating the severity of environmental problems requires seeing their global dimensions. Not until the late 1960s or the early 1970s, after absorbing Barry Commoner's book *The Closing Circle* and the Club of Rome's report *The Limits of Growth*, did Dumont fully embrace ecologism. Before he read those works, he has admitted, his training as an agronomist had him "defending incomplete propositions: fighting against erosion, fighting for the use of manure as fertilizer, and especially fighting against the threat of famine in the Third World" (Dumont 1977: 276). Those fundamental ecological texts of the 1970s made him "understand better the scale of the problems, which concern not only food production, but also threats to energy supplies, to the atmosphere, to aquatic environments, etc." (ibid.) Just before his 1974 presidential campaign, he wrote *L'utopie ou la mort!* to warn that overconsumption of resources, pollution, and disruption of natural cycles of regeneration were leading to the "total and ineluctable collapse of our civilization in the next century" (Dumont 1973: 4). These conclusions meant that ecological problems were orders of magnitude more serious than those targeted by conservation associations or by protest groups such as those that had mobilized to prevent the creation of a new ski area in the Vanoise national park. French environmentalists had too long remained apolitical out of a mistaken belief that the targets of their action were geographically localized and unrelated to larger patterns of resource consumption. Their localism is what sustained their hope of accomplishing their objectives through private benevolent action and occasional protest. Dumont was apparently not impressed that his presidential campaign drew the support of a few sectoral associations, including a group of country homeowners unhappy about the extent of rural development (Allan Michaud 1989: 101). He taught that truly political ecology attacks an entirely different scale of problems: "It is no longer a question of being satisfied with protecting parks and country homes and little birds; we must reinvent our entire civilization." (Dumont 1977: 278)

Political ecology insists that ecological problems are planetary in scope and life-threatening in severity, and that remedying them will require deliberate, organized collective action at both the national and the international level. Thus, to politicize ecologists is to give them the vocation of interacting with the prevailing powers of the community. Partisan electoral activity is indispensable.

More than a perception of crisis, however, is needed if a political program is to be devised. A program sets priorities and chooses appropriate means of achieving its goals. Dumont (1973: 52) complains that Edward Goldsmith (editor of the British journal *The Ecologist* and chief contributor to its influential 1972 special issue *A Blueprint for Survival*) and the authors of *The Limits of Growth* verge on being "apolitical" because of the questions they avoid: "It no longer suffices to proclaim universally that we must 'limit growth' or 'change or disappear.' . . . We will have to modify our entire way of life, all the basic concepts of our civilization—and first of all, its system of appropriation and management of production." Beyond severity, what makes problems "political" is the way that they articulate with the entire range of problems that the community faces. Dumont's complaint against an environmental agenda aimed at "protecting parks . . . and little birds" is that it grossly underestimates the changes that will be necessary in almost every conceivable area of government competence. Dumont speaks of "reinventing our entire 'civilization'" because taking the limits of growth seriously implies making dramatic changes in such policy areas as transportation, economic regulation, international trade, energy policy, employment, and agriculture. "Political," in one of its most ancient senses, means looking at the community as the association of all the organizations in which citizens perform their activities. The polis is where the differing, often competing ends of the various organizations are adjusted in relation to one another. It is the function of "political" activity to give some coherence to the whole: to assign priorities, to impose restrictions, to promote preferred activities. Dumont's complaint against piecemeal environmental activism—saving a park here, prosecuting a polluter there—is that it fails to show how taking ecology seriously would mean rethinking the whole range of public policy objectives.

Finally, making ecology political presupposes ensuring that its program is imbued with a sense of social justice. Justice dictates how the benefits and burdens of community life are properly distributed. Justice provides standards for deciding whether existing distributions of resources are deserved or whether they should be called into question. Where justice has been violated, communities demand that certain principles of rectification be observed. Otherwise, politics is reduced to relations of force, losing its character as the activity through which a people constructs a mutually acceptable common life. Having denounced Goldsmith for failing to take a sufficiently comprehensive view of the nature of our ecological challenges, Dumont (1973: 52) adds that some ecologists also neglect to "analyze who is responsible for the intolerable waste that is leading us to our ruin."

Dumont taught that relations of domination and dependence—quintessentially *political* relations, large-scale relations of power demanding moral evaluation—traverse the entire range of planetary disequilibria regularly denounced by ecologists. A "political" program is not just a "new philosophy of life." It sets priorities and chooses appropriate means as a function of an assessment of moral responsibilities. Taking moral responsibility seriously pushes ecology in the direction of specific ideological commitments.

Exemplifying Politics

Describing the ecology movement's debt to Dumont, Brice Lalonde once quipped "We are all his children" (Bessett 1992: 133). In view of the sometimes stormy relations between the two men, that statement may embarrass Dumont with a wider paternity than he would want to admit. But it is no exaggeration. No one else has been such a consistent and charismatic adviser to ecological activists; no one else has given them such a powerful exemplar of ecological engagement. Surveys have found that Dumont is the most widely read authority among participants in the French ecology movement (Boy 1990: 16; Jacob 1995: 150, 154).

Still, to have much political effect, ideas must be carried over the years by a determined organization. In that regard, Dumont's 1974 campaign did not give sufficient instruction. Shortly after the election, some eighty associations—anti-nuclear groups, practitioners of organic agriculture, members of urban and rural communes, libertarians, environmentalists—made a first attempt to maintain the momentum of Dumont's example. Over the next 10 years, efforts to create a coherent political identity and a unified structure for French ecologism foundered time and again (Sainteny 1991: 13–25). It is significant, therefore, that Dumont stepped in at numerous junctures in the evolution of a green party—not only in the realm of ideas but also in the realm of practical politics. For more than 20 years after his 1974 presidential campaign, he repeatedly lent his prestige, passion, and energy to efforts to ensure that ecology did not fall back into its pre-political ways (Whiteside 1997).

In only one period, from 1986 to 1992, did Dumont's influence seem to wane. In November 1986, party members supported Antoine Waechter's motion affirming the "identity"—i.e., the not especially *leftist* identity—of Les Verts, the French Green Party.[8] Eventually formulated as "neither right nor left," Waechter's strategy of political autonomy signaled a refusal to enter into electoral coalitions with either of the major political formations existing in France (Waechter 1990: 243–244; Brodhag 1990: 41). In the eyes of some observers, it also portended a reorientation of the party's programs in favor of "naturalist" themes such as saving the countryside (Sainteny 1992: 69; Prendiville 1993: 59–60; Cole 1994: 321). Whether this was a fundamental philosophical shift or a change of emphasis for strategic purposes remains debatable (Whiteside 1995). Whatever Waechter's ideological inclinations were, however, the period during which his brand of ecologism was ascendant now appears to have been only a relatively brief interlude in the history of the French ecology movement. In the early 1990s, Dumont's ecologism clearly regained the primacy it had had since the early 1970s.

During Waechter's interlude, Les Verts gained more electoral popularity than ever before—enough to provoke mainstream parties to propose new environmental initiatives and to entice Brice Lalonde to form his own ecology party: Génération Écologie. In 1992 Lalonde succeeded in splitting the ecology vote with Les Verts. This division was the beginning of the end for Lalonde's political ambitions—and for Waechter's strategy of autonomy. In 1993, Les Verts' policy-making convention denied Waechter his plurality, throwing its support instead to Dominique Voynet. Voynet put forth propositions ending the "negative definition of autonomy"—i.e., the negativism implied by Waechter's "neither right nor left" strategy. In true Dumontist fashion, Voynet proclaimed: "I have always felt like a person of the left, but henceforth it is the ecologists who are the carriers of true leftist values." (*Le Monde*, November 16, 1993) Voynet hoped to unite the fractious movement with an "affirmation of the social dimension of ecology, open to trade unionists, feminists, regionalists and Third Worldists" (*Le Monde*, June 25, 1994). At a stormy National Council meeting, René Dumont seized the occasion to assert his perennial theme: that "one cannot be an ecologist without being on the left" (*Libération*, June 27, 1994). With that orientation, Voynet managed to rebuild Les Verts in 2 years. A pre-election agreement with the Socialists in the 1997 legislative elections brought Les Verts their first seats in the National Assembly and earned Dominique Voynet the position of Minister of the Environment (Ariane and Clarisse 1997). Thus, by the late 1990s ecology had unequivocally established its political credentials in France. Dumont's politicizing lessons had come a long way.

René Dumont defines the ideological center of gravity of popular French ecologism. There is no trace of John Muir's mystical love of nature in Dumont's writings, and no discussion of animal rights or of wilderness preservation for its own sake. But his humanism makes Dumont far more than a rational manager of ecosystems. He is moved by social inequality, unmet human needs, and the effects of poverty and ignorance on human potential. He has none of Rousseau's romanticism, but like Rousseau he teaches how the world looks from the perspective of the least advantaged. He sees poverty not as a misfortune but as an effect of exploitation or culpable indifference. He sees environmental destruction the same way. Taking full measure of the planet's finitude and the complexity of its biotic systems, he exposes the scandal of a developed world that ignores the limits of growth to avoid interrupting its binge of consumption.

Among the French, Dumont is the unavoidable reference point for understanding what it means for ecology to become political. The point is not, of course, that he singlehandedly made the state take environmental issues seriously. It is that he powerfully made the case that marginal, incremental environmentalism by no means responded to the ecological threats that humanity faces. Through his example, global resource depletion and environmentally self-defeating methods of production came to be perceived as public concerns. Through his persistent activism, ecologists learned that acquiring political significance requires more than aiming at the development of moral individuality and taking part in sporadic demonstrations.

Dumont has never written an environmental philosophy, but he has *exemplified* political ecology. He has shown that in politics, if not in philosophy, it counts to be a supporter and an adviser, even when the electoral chips are down. Fine distinctions and imaginary hard cases count less than heartbreaking accounts of conditions witnessed in the world's poorest countries. It counts to earn a scientific reputation that makes one unavoidable in official circles and then to use that access to challenge officials to assume higher levels of moral responsibility. Because he has done all these things for decades, René Dumont has, more than any other figure, affected French ecologists' understanding of their political vocation. However, his writings also exemplify the movement's difficulty in elaborating a theoretical critique of the ecological crisis.

From Praxis to Theory: The Ambiguities of Productivism

Ideologically, Dumont's orientation—indeed, the dominant orientation of most of the French ecology movement—is best described not simply as left but rather as left libertarian. Herbert Kitschelt (1990) explains that left libertarians, like the old socialist left, distrust the individualistic values of the marketplace and the inegalitarian society that results from its operation. The leftness of left-libertarian ecologism consists in appeals to equality and democracy, and in a conviction that reconciling humanity to living within ecosystemic limits requires reinforcing social solidarity. In addition, as libertarians, left-libertarian ecologists emphasize "autonomy, democracy, the flourishing of the person and the community," not central planning or bureaucratic management (ibid.: 340).

For anyone attempting to classify "post-materialist" parties in Europe, the "left-libertarian label" is undoubtedly useful. For those interested in green political thought, however, a nagging question remains: Is left-libertarian ecologism theoretically coherent? French commentators have long had their doubts. Marc Abélès (1993: 9) and Dany Trom (1990: 51) both describe greens as ideologically "polymorphic." These political scientists charge that greens throw together an incongruous repertoire of ideas, including populist themes, a critique of parliamentary politics, anti-statism, anti-productivism, catastrophism, scientism, moralizing discourse, Third Worldism, pacifism, and environmental concern. Michel Hastings (1992: 17) accuses greens of offering up only a "syncretic utopia" and describes their discourse as combining neo-traditionalism, libertarianism, and reformism. Jean Jacob concludes his survey of French ecologists with an essay declaring: "The green movement has no 'new ideas.' Rather it is located at the junction of different traditions that it articulates clumsily." (Jacob 1995: 135)

Parts of the green program often appear implausible or even contradictory. How can one oppose the capitalist marketplace and then also reject using institutionalized public power to control it? What sense is there in prioritizing self-realization when awareness of ecological limits is likely to constrict the range of individual activities? Isn't it ideologically garbled to praise the virtues of small, autonomous communities but to characterize the ecological crisis as a global problem demanding global solutions? What grounds are there for thinking that promoting social equality, in and of itself, does anything for the environment? Questions such as these suggest that "left," "libertarian," and "ecological" ideas have not yet been combined in a meaningful way.

French ecologists have always been aware of such questions. To respond in at least a proto-theoretical fashion, they have leaned heavily on one concept: "productivism," which (Alain Lipietz told me in an interview on April 23, 1991) was supposed to be to ecologists what "capitalism" was to Marxists: a way of designating an adversary and a way of rallying greens around a single pole of analysis.[9] Drawing on a number of popular works, I want to explore the meaning of that concept and to expose certain of its theoretical shortcomings.

Like French ecologism in general, the literature decrying productivism thoroughly mixes social and ecological concerns. It does so, however, by problematizing those domains simultaneously rather than reciprocally. That is, it raises social and environmental issues at the same time and in parallel; however, for lack of reflection on the mutual implication of "humanity" and "nature," it leaves a theoretical gap that critics perceive as incoherence.

Ecologism is distinguished from "socialism-liberalism," says Lalonde (1993: 36), by its opposition to productivism: "Society is not made simply to produce, there are activities other than producing to be valued." Productivism is a social orientation toward ever-higher levels of material production. It consists of ideas and modes of organization (e.g., political centralization, the creation of large-scale enterprises, increasingly fine distinctions in the division of labor) that contribute to this goal (Gouget 1985; Journès 1984: 240; Les Verts 1994: 51–56). Productivist societies aim to spur investment, to enlarge industry, and to innovate technologically. These economic strategies allow them to turn out more products. In order for this expansive process to continue indefinitely, productivist societies also have to find consumers ready to absorb their output. An essential mechanism of a productivist society, says Lebreton (1978: 153), is the creation of "false needs." Consumers must be cajoled into always consuming more so that the system's movement never ceases—so that individual investors (or states, in the case of socialized economies) perceive new opportunities, invest more, create new jobs, and push toward a more opulent standard of living. This is what drives our civilization to extract resources and to pollute beyond the finite carrying capacity of Earth's ecosystems. For Dumont (1986: 11), the resulting crisis constitutes the "total defeat of . . . productivist civilization."[10]

A critique of productivism inevitably puts green thinkers at odds with advocates of free-market economic relations. Capitalists develop production processes and goods on the basis of their chances of earning a profit, not on the basis of whether they are environmentally benign. The short-term quest to earn profit encourages firms to exploit resources with little regard for the rate of their natural renewability or for the possibility that, beyond certain thresholds, exploitation of resources may throw whole ecosystems into irreversible decline. In the absence of countervailing regulations, firms have every incentive to externalize unpriced factors of production (e.g., clean air and water that get polluted in the manufacturing process). At a deeper level, the argument that a productivist society stimulates false needs implies that individuals freely making purchases to feed their desires are not the final judges of the desirability of the products they consume. An ecologically responsible society might use criteria relating to products' environmental impact, their recyclability, and their durability to override investors' preferences and consumers' choices in the marketplace.

Yet, as Lalonde insisted, the analysis of "productivism" also sets greens apart from conventional socialists, for it has been a major assertion in leftist politics that unlimited growth creates the conditions for the satisfaction of human needs at higher and higher levels of cultural development. French ecologists usually count themselves as critics not only of the Soviet model of modernization but also of French forms of socialist politics. Lebreton charged that the French Communist Party was actually a conservative force, hardly distinguishable from the right in its "fetishism of growth, its favoring both civilian and military nuclear power" (1978: 308). René Dumont's lifelong search for a "socialism with a human face" forced him to recognize that every existing socialist experiment betrayed his ideals. All had ended up concentrating power in the hands of an oligarchy, restricting free speech, and accepting environmental devastation in the name of rapid industrialization (Dumont 1977: 248–262).

One possible implication of the combined criticism of free-market and socialist economies might be that ecologism is "neither right nor left," as Antoine Waechter thought. But most of the major figures of French ecologism have seen their ideology as a corrected and extended *socialist* vision of a rightly ordered polity. For example, Lebreton (1978: 308) argues that "ecologism advocates a redistributive socialism, based on real needs." Dumont (1977: 285) maintains that greens are "far to the left of the left," and that "we must ceaselessly seek out a less inegalitarian society, one that is more participatory, and more respectful of man and of natural equilibria." Green arguments about the need to constrain decisions about production, consumption, and distribution that are outcomes of unregulated market transactions have more obvious affinities with leftist criticisms of the market economy than with liberal defenses of individualism and private property (Kitschelt 1985: 529).

As used by the leaders of the French ecology movement, "productivism" suffers from ambiguities that diminish its usefulness as a critical concept. It is, for example, very unclear whether productivism refers to a subjective *ethic* of "always more" (and is therefore potentially amenable to a cultural transformation) or whether it consists in a set of social *mechanisms* that condition consciousness, impelling individuals to produce and consume without limit. In the former case, green practice would be largely pedagogical; it would be a matter of changing people's *thinking* such that they became aware of their interdependence with nature, valued it noninstrumentally, and altered their expectations always to increase their material welfare. In the latter case, on the model of Marx's call to abolish private property, dismantling the *mechanisms* that drive production and consumption would be the top priority (Whiteside 1994: 341–344). Describing those mechanisms would be crucial to the ecological project. But many French ecologists used the term "productivism" without attending to such theoretical niceties.

Productivism clearly signifies a commitment to *too much* production—but too much relative to what?

One answer is "relative to the conditions of genuine human self-realization." Ecologists charge that current levels of material production actually impoverish the quality of our lives. An ethic of "more is better" implies that the preferred standards for assessing human well-being are quantitative. Lalonde (1993: 36) observes that ecologists are the only group expressing the need to look beyond productivity as a good in itself and to evaluate the purposes that things serve and the experiential quality of producing them. If maximizing material production were no longer the purpose of labor, it would be possible to imagine restructuring *work* to respond to other standards. It might be possible to make work less alienating by reducing work time, favoring artisanal methods and worker participation in management, and rotating tasks that are tedious but unavoidable (Journès 1984: 219). In the words of Lalonde (1981: 62): "Ecologists want to construct a different type of society, centered much more on human relations, turning around the family, the clan, the neighborhood, the home and household. . . . We do not want to be moving all the time towards an abstract world that tears us away from ourselves, from our roots, from our nature." Communities practicing "convivial" living would arise.

"Conviviality" implies that people could find more space for amicable, egalitarian, noncompetitive relations if they could learn to take back control of their tools and their working time. Conviviality is a matter of freeing up personal relations and allowing more scope for personal development and community involvement. Ideally, society would consist of economically autarchic communities spontaneously integrating themselves into surrounding ecosystems. Dumont (1977: 282) spoke of "elaborating grassroots, self-governing microsocieties, associated among themselves [and] regions building Europe on entirely new grounds." Fournier's call for a "return to the earth" inspired some to abandon cities in favor of rural communes. Such self-sufficient societies were to be utopian islands of ecological sanity. Some inhabitants believed they had founded "liberated spaces" in which inter-personal relations were governed by individual desire and creativity, not by institutional imperatives and uniform rules. "Conviviality" denoted a search for a mode of communal existence rejecting hierarchy in all its forms, allow-ing experimentation, spontaneity, and subjectivity.[11] Thus, on the one hand, "productivism" is the sin of a society that prefers one-dimensional material satisfaction to multi-dimensional forms of human self-realization.

A second answer to the question "too much relative to what?" is "too much production relative to the ability of the natural world to reproduce itself." As Lalonde and Simmonet (1978: 44) explain, the key word for political ecology is not "nature" but "ecosystem," which "designates a rel-atively homogeneous and circumscribed whole of reciprocal relations link-ing living species among themselves and to the world they inhabit." Ecosystemic reproduction requires the completion of complex cycles of energy transfer, purification, growth, and decay. Overproduction interferes with those cycles. According to Lebreton (1978: 317), "true ecologism must make people aware of the limits of the biosphere and, consequently, of their own limits in the domain of consumption." Simmonet (1982: 11–13) begins his study of French ecologism with an explanation of "the organization of nature" understood in terms of the self-stabilizing properties of "eco-systems" embedded one inside another. Since understanding nature this way requires scientific training and empirical research, it is hard to avoid concluding—as Dorst did—that scientists, not ordinary citizens, are best qualified to determine the "limits of the biosphere."

These two interpretations of the underlying premises of productivism tug ecological political theory in different directions. The first direction leads toward moral perfectionism—a conception of perfected *human nature*. Ecologists reject a view of human nature in which the satisfaction of mate-rial desire is equated with happiness. Material abundance, they say, cuts us off from qualitatively superior experiences: convivial relations, creativity in work, responsible self-governance. "It tears us away . . . from our nature," claims Lalonde (cited on p. 76 of Touraine 1983). The second direction leads toward opposition to a "society defined mainly by its capacity to destroy nature"—that is, external, nonhuman, ecosystemic nature. What is the relationship between these two natures?

Too often, French ecologists have assumed that people freed from the grips of productivism would automatically adopt environmentally friendly practices (Journès 1984: 221). That is by no means a matter of course. Even when the planet's population was only a small fraction of what it is today and when humans had only the most minimal of technologies capable of affecting the environment, they managed to alter their surroundings in substantial ways. Upper Palaeolithic hunters left the remains of more than 100,000 horses in Solutré (Goudie 1994: 127). Most of the ancient forests in central France had been felled by the early thirteenth century (Ponting 1991: 121–122). With anything like the present level of population and technological sophistication, it is unimaginable that "autonomous" communities will not impose unwanted environmental effects on one another, even if material production is considerably reduced.

The ambiguity of "productivism" thus hides a painful dilemma for ecological politics: If the "nature" relative to which overproduction is defined is some complex set of ecosystems, there is no assurance that *human* self-realization can be achieved while respecting nature. Protecting *ecosystemic* health may require deference to those whose scientific competence enables them to identify ecologically permissible activities—regardless of whether people feel the need to express their individuality through them. If, on the other hand, productivism is defined relative to the "nature" of the fully realized individual, there is no assurance that the reproductive requirements of various ecosystems will be met. One "nature" demands expertise; the other demands autonomy. This observation is not meant to settle the question of whether French political ecology faces an ultimately insurmountable dilemma. It is meant to suggest that, in many statements, the program of French political ecology is undertheorized. More specific, it shows that "nature" is the contestable concept on which the dilemma hinges. In effect, the popularizers of French ecologism did much to problematize nature and humanity simultaneously, but not reciprocally. This is the truth in critics' charges that ecologists espouse a "syncretic" ideology. French greens insist on raising, at the same time, two problematics, one social (pertaining to human autonomy, conviviality, work, and international justice) and the other natural (pertaining to the survival of all forms of life in Earth's ecosystems.) Exactly how they are intertwined and how conflicts between them would be resolved is left nebulous. This, no doubt, is why every scholar who has surveyed French ecologism has devised a different taxonomy of its various strains.

Alphandéry, Bitoun, and Dupont (1991: 10–11) contend that the main division among ecologists is between futuristic systems theorists and backward-looking partisans of a return to the soil and myths of folk identity. Simmonet (1982: 6) finds two roots of ecologism: the teachings of scientific ecology and the ideas of new social movements that seek to "take *Homo economicus* out of his cramped framework of worker-consumer to consider him as a unique being endowed with desires and a culture." Prendiville (1993: 123–145) proposes a perplexingly eclectic four-part classification: romantics, mystics, and authoritarians; personalist-humanists; modernists (those preoccupied by technology, work, and democracy); and a vast mélange of anarchists, utopian socialists, libertarians, and millenarians.

The most complete survey of ecologism in French is *Les sources de l'écologie politique*, in which Jacob (1995: 16–17) manages to reduce the principal strands of political ecology to three: one originating in "the worrisome teachings of the science of ecology," a "libertarian" perspective that aims "to disengage man from certain constraints (social, economic, and political) in order to guarantee him a maximal and immediate flourishing," and "a third nebula" that "proposes to go back on triumphant anthropocentrism to consecrate a certain form of personalism, which substitutes the modesty of concrete man for the voluntarism of the abstract man of Enlightenment thinkers." What is striking about these categories from a French viewpoint is the unresolved heterogeneity of the ways they frame environmental questions. From the viewpoint of English-speaking ecologism, something else entirely is remarkable: None of the categories of French ecologism turns on, or even addresses, ways of conceiving the value of nature itself. Just when it appears that Jacob may be about to bring up such an issue (by mentioning "triumphant anthropocentrism"), he veers back toward "personalism" and its view of "concrete man." Could this mean that cultural differences run throughout green political thought as it is practiced in the two linguistic communities?

Theoretical Centering

Among English-speaking thinkers, resolving the disparate claims of political ecology has arguably become the primary objective of ecological thought. Certainly any attempt to survey green thought must take account of its multiple contributors: deep ecologists and social ecologists, ecosocialists and ecoanarchists, defenders of animal rights and bioregionalists, to name only a few. But behind this multiplicity stands a more fundamental cleavage: A drive for a philosophically consistent view of nature's value divides ecological thinkers into anthropocentrists and nonanthropocentrists. Both, it should be noted, are species of *centrists*.

"Ecocentric environmentalists," explains Eckersley (1992: 45), are "concerned to protect threatened populations, species, habitats, and ecosystems wherever situated and irrespective of their use value or importance to humans." That broad concern stems from seeing intrinsic value "not just [in] individual living organisms but also [in] ecological entities at different levels of aggregation." Nonanthropocentrists argue that nonhuman things contain *within themselves* features that generate moral significance. They insist that nonhuman things are centers of value deserving respect in their own right, for their own sake. Typically, nonanthropocentrists deplore the "anthropocentrism" of every traditional form of ethical discourse on the ground that it can provide no absolute limits on human uses of nature. "From within the perspective of anthropocentrism," Katz (1997: 105) charges, "humanity believes it is justified in dominating and molding the nonhuman world to its own human purposes."

Other environmental thinkers, however, are unconvinced that a nonanthropocentric ethic is either possible or desirable. Though they condemn the effects of humanity's careless use of the environment, they contend that a well-developed view of human values is sufficient to justify every rationally defensible form of environmental concern. They assume, in Hayward's words (1994: 60), that "if humans become sufficiently enlightened about their own best interests, then they will also pursue the best interests of nonhumans." As a part of their argument, such "moderate" or "enlightened" anthropocentrists may contend that noninstrumental values such as appreciation of beauty or of cultural significance appropriately figure in our assessments of the value of natural things (Sagoff 1988). But those values—like all values—are human-centered at least in the sense that they stem from human judgment and in the sense that they ultimately appeal to the total context of what makes for a good, rich, vibrant *human* existence (Norton 1991: 250). The good is not intrinsic to the nonhuman things themselves.

Debates between those two tendencies are so prevalent in the ecologism of the English-speaking world that every scholar who surveys this field finds it unavoidable. In the most widely referenced source work on green political thought, Andrew Dobson (1995: 5) prepares for an extended analysis of anthropocentrism by arguing that ecocentrism is the feature that most "serves to distinguish ecologism from the other political ideologies." From their nonanthropocentric perspectives, both Robyn Eckersley (1992) and Brian Baxter (1999) make the distinction between ecocentrism and anthropocentrism the organizing principle of their reviews of environmental political theory. Theorists of the opposite persuasion are no less inclined to highlight this dichotomy. To clear the stage for his own "enlightened anthropocentrism," Tim Hayward (1998) decries "two dogmas of ecologism": anti-anthropocentrism and a commitment to the intrinsic value of nonhuman nature. John Barry (1999b: 2) calls ecocentrism a "sacred cow" of green political theory and defends an "anthropocentric moral base" for the ecological transformation of the liberal democratic state. Indeed, philosophers surveying ecologism say that evaluating claims of intrinsic value has become the central task of the whole field environmental ethics (O'Neill 1993: 8; Taylor 1992: 108; Weston 1996: 286; Wells and Lynch 2000: 20). Surely Baxter (1999: 6) is right about English-speaking ecologism when he notes that nonanthropocentric moral claims have achieved such a degree of prominence that "at the very least, devotees of rival moral positions have now to have some argument to weaken or refute the claim of the moral considerability of the nonhuman."

The debate between nonanthropocentrists and anthropocentrists typically[12] problematizes neither nature nor humanity. As much as they reject age-old assumptions of nature's limitless bounty or maintain that humans are no longer unique in their moral considerability, neither side makes reflection on the conceptual interdependency of humanity and nature the focus of its philosophical project. We humans are one thing; nature is another.

Anthropocentrists believe that only humans' well-being interests constitute moral value. We are centered in the sense that our psychic unity gives us the ability to devise a schedule of priorities that dictates the standing of "nature" in relation to our projects. This schedule is independent of what nature is. Tallying up the good served by one course of action or another, anthropocentrists never confuse human interests with those of nonhumans. We know who *we* are. True, nonhuman, "natural" things can be described as having well-being interests of their own. Understanding what allows different natural things to flourish on their own terms can even be essential to the anthropocentric enterprise. Gifford Pinchot's conservationism expressed this aspect of anthropocentric concern well. Prudent forestry management required studying trees' reproductive cycles, diseases, parasites. Such knowledge gets valorized, however, only because humans have an interest in the trees' health in the first place. Humans are humans; trees are trees.

We humans remain unique centers of volition, capable (unlike any nonhuman) of making complex judgments about justice and utility. Prudence dictates that we adjust our social practices to avoid unwanted repercussions on ecosystems, but such adjustments do not differ in principle from those undertaken by human communities in the past to adjust to material shortages or to correct distributional inequities. Anthropocentric ecologism determines only *how much* of nature will be allowed to exist as a function of our best understanding of human interests.

Nonanthropocentrists' assertion that nonhumans have morally considerable interests does not change their understanding of what nature is. Nature is centered in the sense that it consists of things—individual life forms, species, ecosystems—with determinate, self-perpetuating identities. Paying proper respect to natural things requires accepting them *as they are*, independent of our desires and anthropomorphizing sentiments. Whatever exists naturally has a claim to continue existing as it is, on terms consistent with the existence of all other natural things—including humans.

True, humans have values that are unique to them as a species, such as the value of meaningful work or of self-governance. Nonanthropocentrism does not change our understandings of such values. It only circumscribes the area within which we may realize them. Humans have to adjust their uses of nature to avoid disrespecting the values of its constituents. Such adjustments do not differ in principle from those undertaken in the past by communities that extended moral consideration to previously devalued human groups (slaves, women, immigrants). Nonanthropocentrists valorize nature in a new way, without questioning its identity. Trees are trees, and (at least in some circumstances) we must let them be.

In a centered ecological theory, the center is the fixed point. Everything else gets relativized to it. Either humanity is the center and nature is made to accommodate itself to human well-being or nature is the center and humanity has a moral duty to adjust its actions to make room for the flourishing of natural things.

What I will call the *centeredness assumption* that runs through so much English-speaking ecologism obscures another option. There is a possibility that thinking about our ecological predicament might best be developed by avoiding the very habit of "centering" our attention. Rather than focus on how to adjust relations between two presumably distinct entities, one might open up ecological theory by examining how the *identities* of "nature" and "humanity" get constituted—together, reciprocally—in the first place. I call such theory *noncentered*. Now, the early popularizers of French ecologism did not articulate a noncentered theory in this sense. But they did reinforce a rhetorical field in which "human" and "natural" issues are kept constantly intertwined.

What is most characteristic of French ecologism since 1968 is its tendency to absorb diverse values—such as beauty, durability, justice, communal solidarity, and protecting life in all its spontaneous variety—without regard to whether they can be explained from one center or another.

If the early advocates of French political ecology came up short in the area of intellectual synthesis, their eclectic efforts were significant nonetheless because they allowed neither the social nor the environmental side of their agenda ever to disappear from view. The potential unity of the well-being of the planet and the well-being of humanity is almost always assumed. Giving philosophical heft to such an assumption becomes the characteristic task of French ecological *theory*. That is the theme of the rest of this book.

2

Humanizing Nature

Living on continents that, as late as the beginning of twentieth century, still had vast expanses of sparsely settled territory, Americans[1] and Australians[2] have been particularly prone to treating "wilderness" as the epitome of "nature." That treatment is less obvious to the French. When Alexis de Tocqueville visited the verdant Canadian interior in the 1830s, he spoke of going to "le désert." "Le désert" as easily signifies a tangled forest as a parched range of sand dunes. Its biophysical characteristics are secondary, for a desert is defined primarily by what is not there: human habitation. A territory lacking an evident human presence was a great novelty for the French traveler to the New World.

"The most characteristic feature of the geography of France," writes the geographer Philippe Pinchemel (1987: 7), "is the extent to which its landscape has been modified by man's activities. Natural landscapes of forest, heathland and rocks occupy only a minor part of the total area. It is the multicolored mosaic of cultivated fields, meadows, villages and urban areas that predominates, the product of the intense activity of man since ancient times." In a process begun in the neolithic era, agriculture, war, the development of transportation networks, and human settlements of all sorts have transformed virtually every square centimeter of French territory.[3] Long before the common era, ancient Gauls cleared extensive parcels of land; traces of their efforts are still detectable in field patterns today. The Romans began cultivating vineyards around 120 B.C.

If such practices changed France's "nature," they have not been widely deplored, even by ecologists. The French take great pride in the beauty of their much-reworked countryside. In Millet's and Monet's paintings of sun-drenched wheatfields, they see a bucolic environment made bountiful—and beautiful—by human labor. Images of the vineyards of Burgundy, of country roads lined with plane trees, are admired the world over. René Dubos articulates the French cultural preference precisely. He believes that "man feels alien in wilderness." It is something that most of humanity left behind millennia ago when it created permanent settlements sustained by agriculture. For every John Muir, who praised wilderness for its power to regenerate man spiritually, Dubos (1972: 140–141) sees "millions of nature lovers for whom the country means humanized nature." In countries where, until recently, there have been immense tracts of territory in which human impact has been minimal, it is common for environmentalists to become preoccupied with wilderness preservation. For example, the Australian ecocentrist Robyn Eckersley (1992: 28) proposes making a theory's ability to justify "setting aside large tracts of wilderness" a litmus test of its having broken with human-centered philosophies that, in times past, legitimized the unconstrained exploitation of nature. In a culture where few hold up wilderness as the quintessence of nature, however, there is less temptation to dichotomize the human and the natural in this way.

French ecologism develops less by frontally rejecting the idea of "humanizing nature" than by obliquely reinterpreting it. Its epistemological underpinning is a conviction that *understanding the processes* that link "humanity" and "nature" is the first intellectual task of those seeking to create an ecologically responsible society. Virtually unanimously, French ecologists express this conviction in a demand for a renewed conception of "humanism." Ecological humanists strive not so much to propagate a new environmental ethic—one which relocates the center out of which all value radiates—as to promote certain collective, pragmatic shifts in the nature-humanity nexus. In the early years of French ecologism, no thinker went further in developing such theoretical reflections on the nature of "nature" than Serge Moscovici. His works of the 1960s and the 1970s effectively charted out a research program that French ecologists have pursued ever since.

The "Natural Question": The Nexus of "Nature" and "Humanity"

Moscovici's *Essai sur l'histoire humaine de la nature* is a work of astonishing prescience. It was written between 1962 and 1967—well before *The Closing Circle* (1971) and *The Limits of Growth* (1972). In the mid 1960s, a work on political ecology was all the more unexpected from a rising social psychologist. If, as conventionally understood, social psychology studies the processes through which groups "socialize" their members, imparting values and beliefs to them, getting them to adjust their behavior to meet others' expectations, then Moscovici takes a most unorthodox turn. He is critical of the premises of the contemporary human sciences, whose conceptions of "society emphasized the opposition of man to nature." They underestimate society's role of regulating "material forces" and neglect the "creation of productive and scientific faculties" (Moscovici 1994: 367). Moscovici's purpose is to demonstrate that different social groups form out of different relations to the "nature" that their work and technology bring into existence. "Socialization" is not only a "social" process of transmitting norms from person to person; it is reciprocally a matter of our collective engagement with a material world.

The *Essay on the Human History of Nature* begins on a prophetic note. The late twentieth century, says Moscovici (1968: 6–7), must grapple as seriously with "the natural question" as the eighteenth century did with "the political question" and the nineteenth century with "the social question." The political question concerned how to expand political representation beyond the noble class; the social question asked how to manage civil society so as to mitigate economic inequality. The natural question arises out of the unprecedented and pervasive power of modern science. "We are now able, consciously and methodically, to intervene in the biological equilibrium of most plant and animal species, to preserve or destroy them, to change the climate, to modify the cycle of energy transformations. Our geomorphic action no longer knows any limits." (ibid.: 7)

"The natural question" has two meanings, which can be distinguished more clearly that Moscovici usually does. On the one hand, it signifies a growing tendency to problematize social issues around the concept of nature and to make "nature" itself into an ideal. For ages, theorists have worried about how human liberty and political power might be reconciled or about how the workings of society might interfere with the prospects for human self-realization. If ideals of over two millennia of political theorizing were reduced to one-word slogans, they might be expressed as follows: Justice! Virtue! Peace! Liberty! Equality! Moscovici's thesis is that another ideal is becoming increasingly prominent: Nature! Environmentalists, to be sure, but also French students protesting repressive institutions in 1968, regionalists, citizen groups working to improve urban life, animal rights activists, anti-nuclear power campaigners, consumer movement activists and many others turned concern for a world fallen into contempt into a new source of social energy (Moscovici 1990: 7).

The second meaning of "the natural question" points to an epistemological shift that is more radical than any previous change in humanity's relationship to nature. Traditionally, political theorists have regarded "nature" as a *given*. Ancients saw it as an ordered whole whose purposes, once philosophically deciphered, sanctioned rule by those most endowed with nature's preferred capacities. Modern science overturns the teleological conception of nature, and with it, views of the goals of human society. A purposive conception of the cosmos gives way to a mechanistic view of a universe shaped by efficient causes. A nature shorn of intrinsic goals no longer dictates the ideal state of human community. Individuals have to constitute order out of their own will—and master the natural world to make it serve human purposes.

Yet even this enormous intellectual shift leaves intact the supposition that "nature" and "humanity" have independent and definable characteristics. Even as Western thinkers, from the seventeenth century on, gained confidence in the ability of modern science to bring "nature" under human control, they did not cease to assume the givenness of nature. When Thomas Hobbes mocked Aristotelian science, he did so in the name of another "nature" and another "man" that he thought moderns had *finally gotten right*. Knowledge replaced error. There was no awareness that conceptions of "nature" and "humanity" evolve together, such that what must be grasped is not the essential, timeless "nature" of either but rather *the process of their evolution*.

"The natural question," for Moscovici, signals a growing consciousness of this processual linkage. Humanity has transformed the conditions of life—human and nonhuman—beyond the point where their reproduction is assured. Henceforth, we can no longer pretend that "nature" is unalterable in its fundamental characteristics, that it is passively receptive to human designs. Humanity has acquired powers that have turned nature into a *question*, something whose identity is now uncertain. Inevitably, that ability calls our identity into question too. An awareness that we live "on a planetary scale" requires new modes of thought and action—ones that dispense with the assumption that human beings, individually or collectively, are separate and independent from the natural order and superior to it. And since the processes that brought us this awareness were actually at work in every earlier epoch, the natural question sheds a new light on all preceding history. Moscovici devotes the bulk of his *Essay* to plotting out a "history of natural orders"—an account of the co-evolution of society and nature.

Moscovici (1968: 39) defines "nature" as "man with matter." "Matter" designates certain elements and mechanisms interacting according to particular physical laws. "Nature" is the combination of these elements and mechanisms into organized wholes that move through a series of states—"the totality of [elemental] relations concretized into a configuration" (ibid.: 30). Moscovici's thesis (ibid.: 50) is that through a combination of labor, technical knowledge [*savoir faire*], and technology, society makes this configuration appear, while at the same time, this appearance shapes the arrangement of groups in society:

If there is a natural history of biological and social man, it is because matter itself has evolved, and if there is a human history of nature, it is because man—by transforming himself—became able to reconstitute and extend that evolution. . . . [Human nature] resides in this process in which man appropriates and recovers, for himself as an agent, the history of the matter out of which he makes his history. . . . There is no need at all to discover an origin or a permanent end of this: the process alone is what is important.

The nexus of humanity and material configurations has gone through a succession of stages—a series of "states of nature"—that constitute the "human history of nature."

Moscovici refers to the earliest such state as "organic." Beginning in neolithic times and extending to the eve of the Renaissance, he discerns societies in which raw materials such as stone and wool are worked on by artisans. They create objects according to a know-how seen as perfectly adapted to their matter. They often imitate forms and processes that they observe in the world around them. "Nature" and "humanity" are conceived analogously in terms of intelligence and purposiveness (Moscovici 1968: 85–90). The second state of nature is mechanistic. From the sixteenth century on, the growing use of machines has fostered a view of nature that is distinct from a human artisan; it is composed of quantifiable forces (e.g., acceleration, pressure). Matter becomes known not as raw materials with individualized properties but as a homogeneous substance set in motion by energy. Work gets redefined. Skills are separated from labor conceived as force. Work tends to become simply a motive power, interchangeable across production processes. A new "natural category"—engineers—takes precedence over artisans. Experimentation, books, and university teaching become more important than taking on a craft through apprenticeship. Philosophers announce that the technicians' rules, which are based on reproducible observations of measurable phenomena, are "laws of nature" (ibid.: 290–291).

In the nineteenth century we began to enter into a third state, the cybernetic era, in which we see nature as the combination of elements into self-sustaining systems. Knowledge of the properties of these systems allows us to create new forms of matter, not merely to shape or transform what is given. Scientists take precedence over engineers; scientific discovery outpaces social and economic applications. "Artifice" and "nature" are no longer sharply distinguished. Indeed, "natural" and "non-natural" processes are understood through overlapping concepts of information, communication, recombination. In the cybernetic context, human work serves as the "regulator" of systems, seeking to keep them functioning well (ibid.: 98–106).

Labor, technical knowledge, and technology are the main things that hold the nature-society nexus together. "Humanity gives itself a natural foundation," says Moscovici (1968: 53), "when it makes the transfer of its properties to matter, and the transfer of matter's properties to it, the *main object* of its activity. *It erects this foundation not by the fact of working, but by the fact of creating work*." Moscovici emphasizes innovations in work practices—the acquisition of new talents, the introduction of new technologies—as conditions for the emergence of new understandings of what "nature" is. Later (1974: 95) he observed that the visual arts too play a role in this process. Through labor and invention, human beings make the phenomena of the physical world interact in previously unseen ways. For example, when falling water replaced human labor as a force to run mills, a "path was opened to the discovery of gravity" (1968: 55). To cultivate fruit, peasants had to understand the pace of tree growth and the effect of seasons on plants. Early scientists, Moscovici asserts, did little more than translate this practical knowledge into a systematic language.

Labor and technology are not simply *means* through which we transform the world, as if matter passively awaited their transformative power. They *position* humanity in ways that alter our understanding of what nature is. In that sense, humankind creates its own nature.

At the same time, humans are also the subject of nature. The very skills and technological innovations that make nature appear to us in certain ways end up transforming social structures too. The spread of farming and animal husbandry causes hunters to disappear or move on. Those who practice agricultural arts devise elaborate rites and practices to ensure the transmission of their knowledge from generation to generation. Similar processes are at work today. "Tight cooperation between technicians, artists, scientists is indispensable to bring forth the qualities of the physical world, and to interest other collectivities in consolidating their knowledge." (Moscovici 1968: 80) In the process of moving society from one "state of nature" to another, certain groups are the primary movers, the "carriers of invention": the artisans of the ancient world; in the modern era, the mechanical engineer; today, scientists.

In one of his most striking theoretical claims, Moscovici (1968: 119) asserts that the fundamental principle structuring social relations is not wealth or kinship or class but "la division naturelle." According to the principle of natural division, societies' needs for reproduction and innovation encourage the development of distinct groups with different forms of knowledge and skills in relation to the material world. Divisions along lines of sex, age, and occupation stem from a society's mode of interaction with material forces. These differ as communities depend on hunting, or farming, or mechanized production. These "natural categories," as Moscovici names them, are the human medium through which "nature" evolves. The interaction of natural categories, whether ordered in hierarchies or disorganized by competition and conflict, sets in motion the process by which newly dominant skills reveal previously unsuspected properties of matter (ibid.: 155).

Even this human medium has no fixed nature. Human anatomy as we know it is the result of our ancestors' creation and use of artifacts. In his more anthropological work *La société contre nature*, Moscovici traces the development of human characteristics such as the use of tools and communication out of primate societies and argues that humanity's distinction from the apes is the result only of "la division naturelle"—not a fundamental difference of "nature." Thus, " it is illusory to claim that we possess nature as an autonomous, closed or ultimate entity. On the contrary, we elaborate it progressively and restructure it periodically. . . . What we master is a movement, by transforming relations of which we ourselves are a part." (Moscovici 1972: 169–170) Note that when denying that "nature" is "autonomous" or "ultimate," Moscovici is not saying that its features are unknowable or infinitely pliable. To "*master* . . . a movement" suggests that we can achieve greater understanding of the external world and of our own characteristics by studying their processual linkages rather than by examining them separately. The essential question is: what sort of mastery would this improved understanding enable to us undertake?

Moscovici (1968: 24) argues that in our cybernetic era the co-evolution of nature and humanity has reached a decisive point. Humanity is now obliged to "take charge of nature." The natural question challenges mankind to participate in "the government of the natural order." It is important to keep constantly in mind, however, that the "nature" whose "governance" Moscovici envisages is not the "nonhuman" world. We must not imagine governing nature as being like tending a garden or safeguarding a patch of wilderness. Such images reify "governed" and "governor" alike. Nature, to repeat, is the configuration of elements made to emerge by human labor and art, technical knowledge, and technology. And since it is through the creation and dissemination of skills that societies develop their states of nature, "governing nature" takes place through understanding and channeling this process of skill creation.

Political Ecology without a Center

Moscovici (1968: 538–539) calls for a "political technology"—a "science of sciences." Political technology would treat scientific research, the social role of scientists, the methods of scientific training, the economic and social effects of technological innovation as well as the intervention of government bodies in all of these domains as a unifiable field of understanding. The breadth of this charge might eventually bring understanding of the mechanisms of social growth and crisis, which mainstream social sciences cannot do because they treat technological change as exogenous to their theoretical models. Political technology would be "the form of knowledge that would allow people both to direct their collective destiny, and by foreseeing their own evolution, to prompt the emergence of its successive stages" (ibid.: 555).

The idea of a political technology makes it clear that, for Moscovici, ecologism cannot propose to "go back" to nature, to restore some ideal equilibrium in the biosphere as it existed before human intervention. For all practical purposes, "nature" has always been "man plus matter." There is no original condition that we might revisit. In fact, demands that we practice a more "natural" lifestyle, or that we set up nature preserves where ecological processes can continue undisturbed simply reverse the value polarities of societies that have typically defined themselves "against nature." Nature and society are still dichotomized. The difference is only that now we are invited to side with nature "against society." Even doing that, however, is to regard human beings as essentially anti-natural. In fact, by leaving the dichotomy intact, we end up reproducing its structure physically in the world around us. Seeing our ecological challenge as one of creating nature preserves ("protecting wilderness," as many English-speaking ecologists would put it) leads to "dispersing nature in the form of parks disseminated in mechanized surroundings—like scientific objects submitted to observation and control" (Moscovici 1972: 343). Aiming squarely at the rhetoric of a philosophically eclectic ecology movement, Moscovici warns that viewing nature as something fixed and nonhuman actually confuses the choices that societies are now facing. "Ecology will be truly political," suggests Moscovici (1994: 387), "when it admits that what is characteristic of our history is to choose among the states of nature that exist at a given moment."

That is not to say that the ecology movement is wrong to criticize the state of nature as it has been chosen thus far in the cybernetic era. In tandem with activists such as Brice Lalonde and Dominique Simmonet, Moscovici (1978: 54–56) decries the "productivist" bent of modern societies for depleting nonrenewable resources, for generating pollution and waste.[4] But the historicizing framework of his ecologism begins to show how that concept might be employed while resisting its interpretive ambiguity. In a human history of nature, productivism is not defined in relation either to the objectively ascertainable requirements of ecosystemic reproduction or to the timeless, essential needs of the human subject. Productivism describes the orientation of a "society against nature": one that tries, by always increasing production and consumption, to negate its links to nature, to free itself from fears of scarcity and from the destructive elements (Moscovici 1972: 369). Productivism is not merely an ethic of "always more." True to a methodology that makes the mediating activity of work—not the human subject—into the center of "natural" relations, Moscovici emphasizes the inner movement of a productivist society.

That movement depends on refining the division of labor and separating manual and intellectual labor ever more completely (Moscovici 1978: 54). In society's quest for higher levels of material satisfaction, individuals are molded as productivity requires. Efficiency demands the repression of desire for immediate pleasure; global commerce "deterritorializes" the psyche; the machine's need for "normalized" parts works against individual and ethnic self-expression (Moscovici 1976: 102, 113–115). Productivist societies, which make nature appear in the threatening guise of scarcity, divide labor into increasingly narrow categories. Intellectual and manual laborers are separated hierarchically in order to ensure that production proceeds efficiently. Productivist societies "sacralize" meritocracy and technocracy (Moscovici 1974: 97). Distinctions of income, status, and power are all bound together in society's system of "natural division." Thus overcoming inequalities cannot be a matter of attacking the "social question" in isolation from the natural one. Intellectually, too, productivism favors disciplinary divisions. "Each domain—ecological, industrial, scientific, demographic—is supposed to follow its own rules and have its own dynamism, as if there were no relationships between them." (Moscovici 1972: 373)

A "society for nature" would have to redirect this inner movement. It would seek not to arrest growth but to challenge the necessity of technologies and products that are destructive or polluting or that are likely to exacerbate social stratification. More than aiming to protect nature directly, it would problematize the "natural categories" of today's societies: their division of labor, their production of knowledge, the qualities implicit in different technologies. Moscovici commends a "re-enchantment of the world" and the "ensauvagement" of social life. "Ensauvagement" is a neologism meaning "making savage or wild," but it has little to do with preserving wild territories. A reenchanted world dismantles needless prohibitions on individual experimentation, stimulates creativity, and values the diversity of ethnic communities (Moscovici 1976: 123–129). One might best translate "ensauvagement" as "undomestication."

Undoing our domestication implies attacking the forms of socialization that accompany productivism. In a bow to France's alternative groups of the 1970s, Moscovici praises their heterodoxy, their communitarianism, their desire to experience life intensely, their interest in manual arts. Properly understood, their "return to nature" is not backward-looking. It expresses rather a will "to adhere tightly to the ground of society, of nature, and to challenge them violently, and to provoke history" (Moscovici 1974: 32). In 1978, Moscovici urged ecological activists to overcome their distaste for organization, to form local action groups, and to seek connections with heterodox new social movements—urbanists, feminists, organic farmers, regionalists—to protest the practices of productivism. All of them, in fact, share a core characteristic in spite of the diversity in their programs: all of them are "naturalistic." All are protesting how a society dedicated to technical rationality stamps out individuality, how maximizing productivity rides roughshod over a goal of optimizing the conditions for the regeneration of life. Their goal is not a "cult of nature," it is a "praxis of nature" (Moscovici 1978: 104, 110, 130).

In view of Moscovici's conception of nature (1978: 89–97), the program of political technology would also mean being willing to revise the social context of scientific research: challenging its links to the military and to productivist economic practices, imposing respect for life on laboratories and regulating genetic manipulation. Intellectually, an ecological society would produce knowledge in a cross-disciplinary fashion. Moscovici (1990: 17–18) complains that the social sciences still confine environmental issues to the physical and biological sciences, "which are seen as responsible for analyzing the basic structure of these problems and for pronouncing solutions to them." He chides political ecologists for taking up a language "saturated with biophysical metaphors." Unwittingly these intellectual approaches reproduce the traditional dualism of nature and humanity, whereas a new representation of the social is due. "A flexible union of sociology, economics, psychology, history" should be employed in devising new approaches to "the natural question."

This wide-ranging intellectual charge makes Moscovici's work prefigure the research program of political ecology as it has developed in France since the 1970s. It anticipates René Passet's ethically pluralistic environmental economics, just as it suggests the ecological turn in Edgar Morin's quest to unify the human sciences under the sign of a "planetary anthropolitics." In Moscovici, Michel Serres would find a kindred spirit. Just as Serres seeks to limit the authority of science by pluralizing epistemology, Moscovici (1978: 99–100) calls for "a new epistemological path" to take account of "sciences and forms of knowledge that . . . describe reality from two or several mutually exclusive points of view." Moscovici's (1968: 557) proposal for a "political technology" that would give "each member of the social order the opportunity to speak in his own name" is startlingly similar to Bruno Latour's idea of a "parliament of things." Personalist ecologism in toto can be seen as a gloss on Moscovici's (1978: 126) assertion that a "society for and with nature never forgets man, never forgets his nature, which is to be free." Like Denis Duclos, Moscovici (1972: 202) understood the psycho-logic that leads a "society against nature" to fear humanity's animalistic passions and to use that fear as a pretext for imposing far-reaching forms of social control on individual behavior. Moscovici's insistence that the division of labor mediates a society's use of nature reaches fruition in the works of ecosocialists such as André Gorz and Alain Lipietz, with their proposals to put the redistribution of work at the center of the program of political ecology. In sum, Moscovici's great achievement was to think through the implications of "the natural question" to the point where he made his work foreshadow the most important philosophical moves of a rising generation of political ecologists who would problematize the nexus of humanity and nature.

Centered Ecologisms

Seeing Moscovici's writings as anticipating the research program of French ecologism *as a whole* makes it possible to appreciate at a more theoretical level the differences between French and English-speaking ecologism as they have developed since the 1970s. During that period, a vast literature of green theory has appeared in the English-speaking world. As is documented in the previous chapter, a broad swath of that literature addresses the non-anthropocentric claim that nature has intrinsic value. The French, in contrast, barely even took notice of this debate before Luc Ferry's critique of the "new ecological order" in 1992, and Ferry's tendentious charges provoked no major French theorist to take up the defense of the English-speaking thinkers he attacked.

This contrast between French and English-speaking ecologists does not mean that the latter uniformly accept the idea of nature's intrinsic value. Far from it. But it is fair to say that the *controversy* over nature's intrinsic value has become so pervasive that most English-speaking ecologists feel bound to situate themselves, affirmatively or critically, in relation to it. Exposure to the rhetorical field of French ecologism helps us perceive a contestable pattern in these arguments. The impulse to stake out a position regarding the ultimate ground of environmental values drives English-speaking theorists in the direction of centered ecologism. And centering has a price. Anthropocentrism and nonanthropocentrism alike have difficulty assimilating lessons about the intertwining of conceptions of nature and human identity.

Nonanthropocentrists often claim inspiration from the writings of the American conservationist Aldo Leopold (Nash 1967: 6, 63; Callicott 1989: 5–6; Rolston 1988: 160; Katz 1997: xxi). In one of the most influential works of modern environmentalism, *A Sand County Almanac* (1949), Leopold argued that over history we have gradually extended the reach of ethical relations. At first restricted to promoting cooperation primarily between interdependent persons, ethics eventually came to regulate individual conduct in relation to society at large. Now, says Leopold, we are on the threshold of a new extension of ethics: one that makes it pertain to things beyond the human community. Its key principle would be that a thing is "right when it tends to preserve the integrity, stability, and beauty of the biotic community" and "wrong when it tends otherwise" (ibid.: 262) Humanity is finally coming to acknowledge that "soils, waters, plants, and animals" are part of our community (ibid.: 239). Leopold begins to articulate a "land ethic," in which we would recognize that parts of the environment have a "right to continued existence . . . in a natural state" (ibid.: 240). The pivotal claim here is that certain nonhuman things have value *in themselves*.

What exactly is the source of this value? Does Leopold value life in all its forms? Or does "stability" imply that durable patterns of biophysical interaction, not individual lives, generate value by virtue of their homeostatic properties? But then, what do we make of beauty's contribution to the land ethic? Doesn't beauty presuppose the existence of a human observer, and if so, can it be said to be intrinsic to natural things? Moreover, isn't it fair to say that Leopold's own willingness to manage nature means that he did not so much assert its intrinsic value as try to stop the exclusive sway of economic valuation? (Barry 1999b: 124–125) Arguably, Leopold mixed anthropocentric and biocentric reasoning.

Nonanthropocentrists take such ambivalence as a philosophical challenge. In the analytical tradition that characterizes so much Anglo-Saxon philosophizing, ethicists try to bring order and consistency to our jumbled moral intuitions by discovering some single value-conferring property that is the foundation of ethical reasoning.[5] That value-conferring property explains why, and to what degree, the entire range of human activity is morally significant. This property has intrinsic value, in the sense that it is valuable in and of itself. Intrinsic value contrasts to "instrumental" value, value that arises because one thing is "good for" some other end.

Traditionally, philosophers have proposed human beings (or some essential aspect of them, e.g. reason, consciousness) as the prime example of something with intrinsic value. If that value-generating property is unique to humans, then the value of nonrational beings, including all of nature, can only be instrumental.[6]

The innovation proposed by many environmental ethicists is to detect intrinsic value outside of the human subject. Many reject the assertion that reason is the only value-conferring property. If alternative value-conferring properties exist, and if they can be found in nonhuman things, then we must extend respect to those things possessing those properties intrinsically. In a widely noted version of intrinsic value argument, Paul Taylor derives moral significance from the fact that a life form has a condition of well-being. Plants and animals, like human beings, have such a condition peculiar to their species. We can speak meaningfully of a plant's health, of its structural completeness, of its opportunity to deploy all of the functions of which it is capable. Biologists and ecologists can discover the conditions under which various plants and animals flourish. Thus, argues Taylor (1986: 19), "it is possible for us imaginatively to look at the world from their standpoint, to make judgments about what would be a good thing or bad thing to happen to them, and to treat them in such a way as to help or hinder them in their struggle to survive." From the standpoint of each life form, it is intrinsically valuable to be in a flourishing state. Taylor concludes that we have a duty to respect not only human life but every individual life.

Taylor's methodologically individualist interpretation of intrinsic value is, however, atypical in the literature of environmental ethics (Katz 1997: 69). Often what troubles ecologists about human activities is not that they destroy this or that particular nonhuman life. After all, what could be more "natural" than mutual instrumentalization? Individual plants and animals do that all the time to one another in processes that are essential to the flourishing of life throughout an ecosystem. Ecologists worry most when human activities degrade entire ecosystems. Such worry would seem to argue for attaching worth to relationships of complex interdependence, adaptation, and dynamic equilibria—not, or not only, to individual life forms. Many nonanthropocentrists argue that moral considerability inheres in holistic entities (Rolston 1988; Callicott 1989; Katz 1997).

Nonanthropocentric reasoning, however, generates ethical conundrums that call into question its very coherence. For example, once one goes down the path of defending nature's intrinsic value, it is not easy to avoid biocentric egalitarianism—a declaration that every life form has equal worth. According to Taylor, the differentia that, supposedly, give human beings special moral worth always beg the question of what features really *are* morally worthy. True, *we* value reason, and reason is probably crucial to actualizing a *human* life. But other creatures have their distinctive capacities that help them fully actualize *their* lives in relation to their own environment. In the absence of any non-question-begging criteria for moral distinction we must conclude that "animals and plants have a degree . . . of inherent worth equal to that of humans" (Taylor 1986: 152).

Yet biospheric egalitarianism seems ethically absurd. Isn't the notion that natural things have an inherent worth equal to that of human beings inconsistent with the very existence of human life? Acknowledging something's inherent worth means, at a minimum, not deliberately killing it if it poses no direct threat to oneself. Humans who feed and shelter themselves, however, virtually have to kill some plants and animals. Acknowledging *nature's* inherent worth seems to condemn *us*.

Nonanthropocentrists try to avoid that clearly unacceptable conclusion by explaining why certain human activities that entail taking nonhuman life are, nevertheless, consistent with respect for nature. But their explanations flirt with anthropocentrism. Taylor, for example, holds that all species are equally entitled to live according to the same principles. Among equals, conflicting moral claims should be settled in a principled way through priority rules. The most important of these is the principle of self-defense. "The principle of self-defense permits actions that are absolutely required for maintaining the very existence of moral agents. . . ." (Taylor 1986: 265). Like every other species, human beings are justified in taking the lives of other species, but only to fulfill their "basic" interests. And lest that word be used to conjure images of inhumanly austere conditions, Taylor allows that humanity's basic interests include cultural participation. We write down languages, create art, develop technologies. Thus, even if some cultural practices come at the expense of nonhuman lives, they can be morally permissible (ibid.: 281).

Does our cultural need for writing paper supersede the right of pulp-producing trees to exist? Aren't automobiles part of our culture, even if their emissions contribute to global climate change? Taylor cannot mean that just because a human practice has a cultural dimension, it automatically trumps the intrinsic value of nonhuman things. We need ways of assessing the good of cultural practices vis-à-vis harms to nature. But the idea of nature's intrinsic value forbids precisely such calculations of comparative worth; Taylor's biocentric egalitarianism puts the well-being of each lifeform on its own scale. The point, of course, is not that Taylor would want to defend every "cultural" practice. It is that an egalitarian interpretation of nature's intrinsic value has such potentially inhumane implications that even a firm advocate feels compelled to qualify it. In doing so, however, he accords advantages to humans that seem to give them special entitlements in relation to everything else that is natural.

Making nonanthropocentrism more holistic does not solve the problem. Consider the reasoning of Lawrence Johnson. Johnson believes that he persuasively counters the egalitarian logic imbedded in the notion of nature's intrinsic value. He argues that any living system that can be said to have "integrated effective functioning of [its] self as a whole" carries some moral considerability (1991: 142). Since this is true of species and ecosystems as well as of organisms, we must to some degree respect them all. Johnson proposes that we "give *due* respect to all the interests of all beings that have interests, in proportion to their interests" (ibid.: 118). His ecological holism grants that all life systems have morally considerable interests—but that some lives count for more than others. He proportions moral consideration to "complexity, diversity, balance, organic unity, [and] integrity," which give living beings their identities (ibid.: 56).

On closer analysis, Johnson's ethic betrays suspiciously anthropocentric tendencies. Though it is plausible to say that different creatures have different types of interests, why should that justify different levels of moral consideration in cases where humans and nonhumans conflict over the *same* interest? All life forms, ourselves included, have an extremely powerful interest—arguably, a pre-eminent interest—in preserving their own lives. Doesn't that interest deserve equal consideration, no matter where it occurs—and isn't Johnson being anthropocentric when he denies such a conclusion? He defends his stance by arguing that people deserve more consideration because, in comparison to other creatures, their interests are more numerous and complex, and they have greater awareness of them (Johnson 1991: 186–188).

This argument proves far too much. What chance does a snail darter have against a creature whose technological sophistication allows it to build a staggeringly complex electric power network that lights theater stages and runs kidney dialysis machines? If special cognitive skills and functional dexterity increase the moral weight of human interests, then whenever human interests compete with nonhuman ones, humans will win. Anthropocentrism peeks through again. The disclosure of anthropocentric tendencies in the works of dedicated nonanthropocentrists lends credence to Tim Hayward's view that "as long as the valuer is human, the very selection of criteria of value will be limited by this" (1998: 51).[7]

Perhaps we have no choice but to own up to anthropocentrism while making it more environmentally sensitive. A number of English-speaking ecologists do precisely this. They break with what Bryan Norton (1984: 134) labels the "strong anthropocentrism" of the past: one in which "all value . . . is explained by reference to satisfactions of felt preferences of human individuals." *That* anthropocentrism makes existing human preferences morally dispositive. It sanctions the unconstrained instrumentalization of nature wherever people have not already discovered a reason to value it in its undisturbed state. On the other hand, "weak" (Norton 1984), "reflexive" (Barry 1999b: 39), or "moderate" (Hayward 1994: 59) anthropocentrism strives to show that human interests, rightly understood, suffice to ground the very sorts of environmental concern that motivate nonanthropocentrists.

Most straightforwardly, one can make such a case by highlighting the human interests directly served in each area of environmental concern: health interests, economic interests, aesthetic or spiritual interests. Our duty to protect those parts of nature that serve those interests are really duties not to harm other people and to treat them justly. No special environmental ethic is necessary (McCloskey 1983: 33; Passmore 1974: 187).

Nonanthropocentrists object that such reasoning is too weak to justify many plausible demands for preserving nature. Katz (1996: 315) charges that any ethic grounded in "human desires, interest, or experiences" will fail to specify "the source of moral obligations to protect the environment," because those human phenomena "are only contingently related to the continued existence of wild nature as such." Defending wilderness in terms of aesthetic or recreational pleasure fails to explain why the entertainment we feel in a theme park's artificial jungle so inadequately substitutes for experiencing nature's power, variety, and sheer otherness. Anthropocentrism also has difficulty explaining why there is anything wrong with cruelty to animals or with allowing the extinction of an ecologically redundant species.

Finally, anthropocentrism seems to incline theorists to minimize the significance of ecological interconnectedness. H. J. McCloskey, for example, finds Barry Commoner's ecological rule of thumb—"Everything is connected to everything else"—unsuitably broad if it is used to justify an a priori caution about intervening in nature (McCloskey 1983: 52). Only where scientific evidence proves that interconnected environmental phenomena detrimentally affect human goods is caution mandatory. Nonanthropocentrists suspect that those who reason from the human good alone are all too quick to belittle evidence of interconnectedness. Anthropocentrists such as McCloskey do so out of their own a priori commitment. As debates over global warming have shown, some people are determined to maximize the scope of human freedom—even if liberty can be secured only by denying reality.

Some green theorists maintain that *enlightened* anthropocentrism remedies such defects. Tim Hayward has developed one of the most sophisticated versions of this perspective. To count as enlightened, he argues, one
must first come to appreciate not merely that one's interests are bound up
with those of others in a web of interdependency. In addition, one must
realize that "the very substance of one's interests is at least in part the product of one's interdependence with others" (Hayward 1998: 67). The realization that one's interests are in part socially constituted makes possible a
critical reevaluation of them. Enlightened individuals, in recognition of their
interdependence and equality, agree to subject the pursuit of their interests
to constraints of justice decided deliberatively with others (ibid.: 77, 98).

Allegedly, enlightened anthropocentrism would more consistently defend
the environment than the straightforward version. This is because the delib-
erative-critical process of enlightening people's interests would make them
more cautious about causing environmental harms whose existence pre-
enlightened self-interested individuals might have been tempted to deny
(Hayward 1998: 3, 106). Presumably even somewhat speculative harms
that issue from ecological interconnectedness or that might affect mainly
future generations become ethically compelling in the context of heightened social solidarity.

Moreover, enlightened anthropocentrism defends the environment in
ways that nonanthropocentrists have argued are uniquely available to an
ethic of nature's intrinsic value. For instance, understanding the good of
bodily integrity, Hayward's enlightened individual extends compassion
to other creatures. Those creatures' efforts to maintain their integrity—to
maintain their bodies in being—express a degree of freedom not wholly
dissimilar to our own. This freedom "provides a prima facie reason for
respect for all beings" (Hayward 1998: 137). Enlightened anthropocentrism thus forbids the "wanton" destruction of other species. Finally,
respect for other species "on the basis of *their* species capacities" (ibid.:
146) undoubtedly requires preserving the ecological context in which
those species capacities alone get fully deployed. There are no spotted
owls without old-growth forests for them to live in. Thus the enlightened
anthropocentrist can favor wilderness preservation on noncontingent
grounds as well as on aesthetic and recreational ones.

Alternatives to the Centeredness Assumption

This brief rehearsal of debates between English-speaking nonanthropo-centrists and anthropocentrists only hints at the range of positions available to those embroiled in the question of how and to what extent humanity must value nature. Additional twists and turns get examined in coming chapters. But this presentation of the debate already suggests how far French non-centered ecologism stands outside the field of English-speaking ecologism. Moscovici sidesteps both centered ecologisms when he holds that *neither* human interests nor nature exist independent of one another. There is not a nature "out there" either to be perceived as a function of human interests or to be valued in and of itself. Correlatively, human interests do not have an existence separate from the skills and technologies with which we interact with the material world. Nature and society "institute" each other, maintains Moscovici, in a process whose complexity can only be grasped historically. Once one has encountered "the natural question" as Moscovici frames it, what stands out about centered debates is less how they contradict one another than *how they fall silent at exactly the same points.*

It is characteristic of centered theory, first, not to problematize nature in anything like Moscovici's historicizing sense. Expressions such as "the biotic community" (Leopold) or "holistic entities" (Johnson) or even "animals and plants" (Taylor) take for granted the identity of value-generating centers. From the viewpoint of a human history of nature, Callicott's (1989: 22) assertion that the conceptual foundation of the land ethic is "clearly the body of empirical experience and theory which is summed up in the word ecology" stuns by its confidence in the fixity of its object. But moderate anthropocentrism is only slightly different. Hayward's (1998: 137) talk of "reverence for nature as a whole" is no more likely than nonanthropo-centrism to make us linger over the historically situated processes of scientific research, technological innovation and social differentiation that feed into contemporary conceptions of "nature's" complex functioning. Even less does either centered ecologism invite us to step back from characterizations of nature favored in contemporary discourse while striving to grasp a process of change that might supersede those very characterizations.

It is no accident that centered ecologisms take for granted the identity of natural things: they *must* do so in order for things to carry out their appointed philosophical functions. Johnson's respect for "self-identical" entities reflects this imperative for nonanthropocentric theory. Certainly we cannot respect *nature's* intrinsic value if "natural" characteristics are a projection of *human* identity. And to elicit our moral concern, an entity's characteristics must be stable enough that we can recognize them as belonging to that thing *as such*, no matter when our perception of them first occurs. And so nonanthropocentrists typically depend on scientific descriptions of natural entities, without further inquiry into the larger circumstances that yielded them.

Anthropocentrism has a similar dependency. To have an interest is to have an interest in some*thing*. An interest is not merely the projection of will into a void. Human interests pertain to states of affairs in the world. Those states, in order to count as interest-activating conditions, must be characterized credibly as actually or potentially existing, independent of our interests. Thus, anthropocentric ecologists too continually reference scientific knowledge without systematically examining the conditions under which that knowledge took shape (Evernden 1992: 101).

Making theory noncentered requires a partial relativization of scientific knowledge. Moscovici's (1978: 16) characterization of political ecology as an "anthropological" movement speaks to this difference between French ecologism and centered theories. Like Edgar Morin, Bruno Latour, Michel Serres, Jean-Paul Deléage, and Denis Duclos, Moscovici relativizes our understanding of nature by making it undergo an anthropological moment.

In an anthropological moment, the theorist temporarily suspends belief in the truth of scientifically established information in order to consider our knowledge of nature as a *human* phenomenon: one in which understanding follows not just from impartial curiosity, painstaking collection of data, and rigorous methods of analysis but also (as Moscovici teaches) from socially developed needs and technologically mediated competences. Other French ecologists highlight how rituals and taboos, cultural preferences and prejudices, organized interests and political ambitions affect understandings of nature. None of this is intended to invalidate scientific information per se. Certainly, Moscovici's (1968: 524) awareness of dangers of climate change and the destruction of species depends as much on scientific evidence as that of any English-speaking ecologist. An anthropological view of the development of knowledge, Bruno Latour (1999a: 79) will later claim, actually *makes science more realistic*. The anthropologist's insistence on studying the mundane details of knowledge-building while suspending belief in scientists' own tales of "objectivity" has an important normative effect: by exposing contestable concepts and unacknowledged contingencies in the processes of research, it opens some of our most common assumptions about mastering nature to discussion, criticism and reform. Theories that take nature for granted foreclose this possibility.

More surprisingly, in comparison to noncentered ecologism, centered ones undertheorize human subjectivity, too. "When people denounce the harm associated with a technology or science," notes Moscovici (1972: 330), "they often seem to take the viewpoint of Man—capital 'M'—that is, they claim to look at things from a universal point of view." That "pure fantasy" (as Moscovici calls it) is evident when a nonanthropocentrist portrays "human beings as contextual beings, whose conditions of flourishing depend on their relationship with a wider context" (Baxter 1999: 108). It is barely distinguishable from the anthropocentric argument for enlightened self-interest, which "reveals that the self is part of a greater whole and related to all other selves" (Hayward 1998: 88). Talk of "human beings" simpliciter allows theorists to dodge inquiry into our psychic complexities.

Moscovici offers one way of delving into that complexity. He holds that our characteristics as psychological agents are always tied to a historically bounded state of nature. *Society against Nature* combines ethology and anthropology to show that what are often assumed to be universal human behavioral characteristics and competences emerged gradually out of collective efforts to control the material conditions of the community's existence. Movement from hunting and gathering societies to ones dependent on farming and animal husbandry, for example, not only brought out new properties in plants and animals but also valorized different instincts and required new forms of self-discipline in human subjects (Moscovici 1972: 364–366). Attributes of our psychic "nature" are inseparable from understandings about external nature that prevail in a given society at a given time. If Moscovici is even close to being right about the human history of nature, then centered views of the personality systematically deprive green theory of this level of reflexivity.

One last distinction is important to grasping what makes Moscovici's approach noncentered. Tracing the processual connection of society and nature is not tantamount to declaring the "social creation of nature."

In a work of that title, Neil Evernden reviews the ways that thinkers from Plato to Aldo Leopold have variously conceived "nature": as the unchangeable essence of things; as the embodiment of ethical norms; as matter in motion. "A modern world view," Evernden contends (1992: 28), requires "a fundamental separation of 'human' and 'nature'." For scientists to study "nature" objectively, every trace of human subjectivity, every notion of inner purpose, had to be withheld from their observations. The dualism that results—mechanical, purposeless nature on the one hand, observing, purposeful man on the other—harbors, however, a fateful irony. After so successfully explaining physical phenomena, scientists applied their methods to life—ultimately to human life. In that case, humanity becomes a subcategory of nature: determined in its actions, purposeless. Consciousness disappears into neurophysiology. How can that be? Scientific reductionism presupposes a cultured, intending consciousness to carry it out.

Paradoxically, however, those who put purposiveness first get caught in the same dilemma. In one of his most perceptive arguments, Evernden shows that Aldo Leopold's "land ethic" and its philosophical siblings face two unsatisfactory alternatives. Some attempt, implausibly, to extend rights to creatures that lack features of agency presupposed by a notion of rights (e.g., the ability to give consent). Since these creatures cannot speak for themselves, their well-being has to be ascertained by other means—namely, by scientific methods of inquiry. Other theorists bring "nature" within the ambit of moral agency by seeing it as an extended human self. In that case, human beings become caretakers for all life and "nature" loses its independence. Whether they see nature objectively as an order independent of humanity or subjectively in terms drawn from human experience, theorists are stuck in an "underlying dualism" (Evernden 1992: 100–102).

To extricate ourselves, Evernden argues (ibid.: 99), we must realize that "this perpetual oscillation between the domains of nature and culture, arises from a fundamental error." "The dualism," Evernden asserts, "cannot actually be resolved, *because it never existed*. The dualism we fret over only exists because of our own decisions. . . . One might even say that there is no 'nature,' and there never has been." Evernden recommends that we abandon "nature" once and for all because nature is *only* a social creation. Socially constructed and reconstructed a thousand times, it is hopelessly caught up in metaphysical dualism. If we are to become more solicitous of the world around us, we should look to what we "encounter before we discover 'nature'"—before whatever we encounter gets dissected and reassembled and mythologized in our cultures. Taking a cue from phenomenology, Evernden sets forth a conception of "the ultrahuman." The direct apprehension of a phenomenon "accepted *in its full individuality*, as a unique and astonishing *event*," may evoke a quasi-religious sense of awe. A person who gets wonder-struck in this way could be "a more benign creature," concludes Evernden (ibid.: 124).

The dualizing rhetorical field of English-speaking ecologism makes itself felt in even this resolutely anti-dualist argument. First, Evernden criticizes the ahistorical realism of intrinsic value conceptions of "nature" by upending them: if nature is not real, it is socially constructed. But, contrary to his intentions, that move seems to leave the world (whatever one wants to call it) entirely open to human technological manipulation. Thus, he goes in search of something with its own *fundamental identity* that can serve as a limit to such designs. He finds the "ultrahuman"—and, as in so much English-speaking ecologism, the "ultrahuman" is virtually synonymous with "wildness" (Evernden 1992: 121). Wildness is suspiciously close to the most common meaning of "nature": "everything which is not human . . . , [that which is] 'otherness' to humanity" (Soper 1995: 108). Following Evernden's historical approach, however, there is little reason to think that the "ultrahuman" is truly fundamental. As Roderick Nash maintains (1967: xiv), a notion of "wildness" makes sense only for those whose culture contains a conception of "controlled and uncontrolled nature." If "nature" is a social creation, so too is "the wild."

That would lead us back to an unsatisfying constructivism. Social constructivism, in that case, is only a variant of the "subjective" component of the traditional dualism. The constructivist attributes agency to society rather than to the individual cogito. But, like the cogito, society is ultimately free in its own constructive activities. Society constructs "nature," society constructs "the wild." Instead of a nature-culture dualism, we are left with a dualism of the wild and the social. Evernden gets into these difficulties because his motivation is the same as that of both anthropocentrists and nonanthropocentrists: he is looking for some identifiable *thing* that can be the locus of ecological concern

Moscovici, on the other hand, leads French ecologists into a different set of strategies for dissolving dualism. Like the constructivists, he investigates cultural-technological processes involved in making nature appear. But a residual realism is always present in his work. Realism manifests itself in a determination never to let go of the results of empirical research in the natural and social sciences, even as it historicizes them, probes them for hidden value assumptions, and looks for influences that cross from one domain of knowledge to another.

In sum, ecologism becomes noncentered when three ideas become integral to its perspective. First, "nature" does not have value-radiating or interest-activating centers. It consists of things whose identities depend on the social and scientific processes through which they became characterized. Second, the psyche is understood to harbor a profusion of impulses, emotions, and aspirations that make it impossible to refer to "humanity" as a unified essence. Third, humanity and nature are not portrayed as separate entities that can make claims on each other. They are conceptually interdependent through and through.

The Question Concerning Humanism

Moscovici would find centered ecologisms inadequate because they exemplify what is most problematic in the discourse of traditional "humanists." In their penchant for textual analysis and ethical speculation, centered ecologisms tend to disregard the "exterior world"—the place of work, science and technological change in the development of social formations and ideas (Moscovici 1968: 505). Ingenious arguments about interests and values largely displace discussion of different conceptions of nature formed by scientific inquiry. Debates for and against nature's intrinsic value typically downplay investigation of the historicity of nature—especially of its relation to changes in technology and the division of labor. Theories of the relationship between ideas and social structures fall outside the scope of philosophical interest. Guided by the centeredness assumption, many green thinkers regard the locus of value, not the identity of humanity or nature, as what is most problematic for environmental ethics. It is not far-fetched to imagine that Moscovici might even connect the attributes of the ethicists' subject—its attachment to rule-following, its desire for generalization, its expectation that moral intuitions be tested against potential counterexamples almost like scientific hypotheses—to a society in which science and high technology so crucially mediate relations with our material surroundings.

Perhaps the most troubling question to be asked of a noncentered ecologism is whether it ends up once again permitting humanity's unconstrained right to change its surroundings as a function of its own desires (Köhler 1983: 28–29). Is it really possible to accept "nature" and "humanity" as reciprocally determining without legitimizing technologies and work practices that end up eliminating an animal species or turning a beautiful canyon into a rock quarry? In fact, doesn't Moscovici's call for "the government of the natural order" represent the ultimate extension of a Cartesian project of bringing nature entirely under human mastery? As will become apparent in the course of this study, such questions can aptly be addressed to almost all of the French ecologists, not just to Moscovici. Thus, we should defer an answer until we have more fully examined the variety of ways that they reciprocally problematize "humanity" and "nature." Here I suggest only that, in preparing to respond that question, we must concentrate our attention on the concept of "humanism"—a concept constantly evoked in a positive sense by French ecologists but often ignored or regarded with suspicion by English-speaking ones.

In the rhetorical field of English-speaking ecologism, "humanism" frequently gets confounded with anthropocentrism. David Ehrenfeld's *The Arrogance of Humanism* makes the classic case. Ehrenfeld maintains that "humanism" is the dominant religion of our era. The central tenet of this faith is that "all problems are soluble by people" (Ehrenfeld 1978: 16). Humanity seeks to "engineer" its own future by scientifically studying phenomena, then devising technological remedies for social problems. At the head of Ehrenfeld's list of things that are lost when we approach the world humanistically is *wilderness*, which is not any particular species or habitat type, but a higher class of life form with its own nobility derived from its complete independence of human beings" (ibid.: 255). The deep ecologists Bill Devall and George Sessions (1985: 54, 111, 118) similarly combine a critique of humanism with a desire to protect "vast areas free as untrammeled wild places." In fact, critical assessments of "humanism" are a commonplace of the nonanthropocentric literature.[8] It is not unknown for English-speaking ecologist proudly to adopt the label humanist. Yet then it means "enlightened anthropocentrism" (see Hayward 1998: 73–84)—i.e., a version of centered ecologism.

Moscovici foreshadows a noncentered understanding of humanism. As a sociologist of knowledge, he has argued (1968: 500–512) that "humanism" emerged in the world of Greco-Roman antiquity, as a result of "la division naturelle." It is part and parcel of a "society against nature," one that strictly divides intellectual from manual labor, and that needs an elite educated in the "liberal arts" in order to master the rhetorical means of maintaining social order. Unlike English-speaking ecologists, however, Moscovici's solution (ibid.: 557) does not implicitly separate humanity from nature:

> . . . the government of nature and the government of society are both—contrary to what has usually been believed—the government of men. Those who act so that the government of society may assure justice and freedom are invariably led to devote themselves to the government of nature. . . . Thus, in order to resolve the social question—whose solutions and programs were stipulated in the nineteenth century—each collective force is obliged to include the natural question among its concerns.

Political technology, as Moscovici conceives it (1976: 111), signals a new moment in mankind's *consciousness* of its role in instituting natural processes and regulating its own structure in relation to them. Embracing this level of understanding no more entails abolishing humanism than rejecting science. It does imply reorienting humanistic studies so that they no longer behave as if the human spirit realized itself in works of reflection and imagination conceived apart from society's relationship to its physical environment. Political technology calls for scientists and humanists to "integrate themselves into the history of their nature" (Moscovici 1968: 555).

Conventionally, "humanizing nature" has implied that there is some knowable nonhuman world out there ("nature"), which we then alter ("humanize") to make it better conform to our needs and wishes. The principal contributors to French ecologism generally have something different in mind. For them, "humanizing nature" means that the *concepts* of nature and humanity are bound together in historical-cultural processes (variously described by various theorists), such that what nature *is* can be understood only in relation to human practices, hopes, and fears—and vice versa.

Nature is never simply "out there," to be encountered in an unadulterated form. Neither is humanity "in here," the essence of an autonomous, reflecting subject. Nature's nature is inseparable from organized human practices. Human nature is inseparable from the influences of a bio-physical reality.

This awareness achieves expression in demands for a renatured "humanism" that are found almost universally among French ecologists (Laurens 1991). The regulationist economist Alain Lipietz (1989: 60; 1993a: 22–23, 66) declares unequivocally that "ecologism is a humanism" because it understands that "man and nature form a whole, man is part of nature, and nature is irresistibly humanized." The systems ecologist Edgar Morin (1993: 166) describes ecological politics as a matter of "civilizing civilization" or the "continuation of hominisation." The personalist Jean-Marie Pelt (1977: 263) hypothesizes that the "ecological current" could well be "the harbinger of the discovery of new world in which man, by enlarging the field of application of the traditional values of humanism to nature and the cosmos, would reconcile himself at the same time with himself." Denis Duclos (1996: 306) argues that "ecology can be a plausible vehicle for a politics of the sentiments" that is "humanistic in the concrete sense—'human' means that which emerges from the humus." Similarly explicit vows of humanism can easily be found in the work of Félix Guattari (1989: 49), Joël de Rosnay (1995: 321), Jean-Paul Deléage (1993a), Phillippe Saint Marc (1971: 11–12), Christian Brodhag (1994: 84), and the green-sympathizing authors of *L'Équivoque écologique* (Alphandéry et al. 1991: 271).

I have deliberately belabored the evidence here for two reasons. First, however ambiguous the term "humanism" might be, its explicit and nearly unanimous endorsement in French ecological circles alerts us to the presence of an ethos distinct from that prevailing among English-speaking ecologists. To understand French ecologism properly, it is vital to correct the misperception created by Luc Ferry's *The New Ecological Order*, which accuses ecologists of anti-humanism. When Ferry criticizes English-speaking authors, he at least chooses targets who acknowledge the tension between their views and some version of "humanistic" ideals. But by confounding English-speaking and French ecologism, he misses what gives his compatriots' thinking its distinctive character. In French ecologism, debates take place not *between* nonanthropocentrists and anthropocentrists but rather *through* various notions of humanism.

The *variety* of humanisms is crucial here. The second thing to emphasize is that ecologists who appeal to "humanism" have an extraordinarily rich heritage to draw upon—a heritage rich enough to fuel controversy quite apart from any pretensions to do away with humanism entirely. Certainly humanism *can* be rationalistic and anthropocentric. Ferry and other liberals regard the essential, law-giving power of reason as constitutive of our humanity. This humanism sees as its forebears thinkers such as Descartes and Bacon, Enlightenment philosophers such as Kant and Condorcet, and social reformers such as Saint Simon and Bentham. Those thinkers represent, however, only one line of humanist descent.

Ecologism as a Skeptical Humanism

Skeptical humanism has an equally distinguished lineage. Moscovici (1974: 28, 64) is attracted to Montaigne, a writer who still jars our comfortable certainties about what is "civilized" and what is "savage"—just as ecologists challenge the distinction between "humanity" and "nature." Pascal, a devoted reader of Montaigne, is a favorite referent of Edgar Morin, Michel Serres, and Jean-Marie Pelt. Jean-Jacques Rousseau has been called the "spiritual father" of contemporary political ecology movements (Schneider 1978).[9] A distinction formulated by the semiologist Roland Barthes helps express the core of the humanistic tradition to which these thinkers belong. The tradition mentioned in the preceding paragraph he would call "classic humanism." It "postulates that in scratching the history of men a little, the relativity of their institutions or the superficial diversity of their skins . . . one very quickly reaches the solid rock of universal human nature. Progressive humanism, on the contrary, must always remember to reverse the terms of this very old imposture, constantly to scour nature, its 'laws' and its 'limits' in order to discover History there, and at last to establish Nature itself as historical." (Barthes, quoted on p. 27 of Evernden 1992)

The centeredness assumption that runs through English-speaking environmental ethicists gives many of its contributors affinities with the intellectual model of classic humanism. For examples of moral theorizing they turn to Aristotle and Kant and Bentham and Rawls. They seek a systematic ethics of interrelated rights and duties, models of impartial decision making, and ahistorical rules of justice. Once they have located the center of intrinsic value in nature or man, they have an extraordinary faith in the power of reason to devise an internally consistent set of rules and measures that will allow us to give every being its due. It is this similarity that lets anthropocentrists charge that, since human beings will necessarily end up making decisions about what constitutes the welfare of nonhumans, every ethic is at bottom anthropocentric. The very intellectual complexity of a nonanthropocentric ethic can be taken as evidence that reason—"the solid rock of human nature"—is the true measure of moral worth, not qualities inherent in natural things.

But what if the drive towards ethical completion itself were part of the problem? In her late works, Judith Shklar argued that overconfidence in the "normal [rationalistic] model of justice" makes people insensitive to wrongs ignored or prettied up by that model. She contended that many of our most deeply felt ethical convictions—our endorsement of tolerance, social equality, democracy, our awareness of victimhood in places where it has not been seen before—arose less by rational systematizers measuring social practices against abstract philosophical standards than by skeptical questioning (Whiteside 1999). Skeptical humanists point to the gaps in every philosophical system; they demonstrate the potential for abuse that occurs when an intellectual perspective is stretched to cover the totality of experience.

Skeptics strive for ethical effects by playing one type of knowledge off against another. They listen attentively to claims of those whose views have been devalued by prevailing conceptions of what is right or "civilized." But when skeptics catch others out in contradiction or ethical deficiency, they do not automatically turn to formulating a new, more complete ethic. They cultivate an appreciation of the irreducible diversity of values and throw them in the face of the forces of commensuration and uniformity.

The only quibble that ecologists might have with Barthes's description of this second humanism is his labeling it "progressive." "Progress" is a term too bound up with notions of material abundance and normalization of behavior to be allowed to name a perspective that often challenges those very ideals (Moscovici 1976: 102–103). It was not an intention to foster "progress" per se but a skeptical turn of mind that made Rousseau into a critic of the "old impostures." Rousseau charged that science and progress worked to man's moral detriment, distancing him from a more "natural" life in which needs and desires were in equilibrium. In the *Discourse on the Origins of Inequality*, he cast doubt on commonly accepted ideas that social inequalities reflected the "natural" distribution of human abilities and that vices such as hypocrisy and indifference to the suffering of others were "natural" traits of the human ego. His speculative anthropology suggested that those traits arose historically, as societies sharpened the inequalities made possible by the introduction of agricultural technologies and the institution of property.

Tzvetan Todorov reminds us that in France the humanist tradition begins with Montaigne. Montaigne saw so much of man's stupidity that he wondered whether human achievements really exceeded those of some animals. Even as Montaigne marked off a domain of "purely human" activities (as opposed to those relating to God), he never forgot, says Todorov (1998: 15), that "human life is destined to remain 'an imperfect garden.'" One recalls Montaigne's remarks on the unchristian savagery of the Spanish conquistadors in the New World and his ironic musings on the relative civility of cannibals. In an elaborate metaphor, Montaigne (1967: 99) notes that we call fruits "wild" [*sauvage*] when "nature" appears to produce them by itself; in truth they are fruits that "we have altered by own our artifice" in order "to accommodate the pleasure of our corrupted tastes." Montaigne's "nature" is inseparable from social convention; like Moscovici's, it has a human history.

Like classic humanists, skeptical humanists are profoundly concerned with human welfare and they believe that the human spirit is free. But they are also moved by a sense that human reason is changeable and corruptible. They lament that we so often mistake unthinking adherence to custom for freedom, even when customs are cruel and confining. Reason is too feeble an instrument to serve exclusively as a guide to ethical conduct—to be the *center* of moral significance. Skeptics criticize what dehumanizes our existence: vain pretensions to social superiority; fanaticism; disregard of the spiritual consequences of technical innovations; philistine outlooks that flatten the moral terrain in the name of utility.

That, I believe, is the general thrust of French ecologism. French ecologists argue that our contemporary world is *humanistically* impoverished because it fails to recognize—or overrides—the great diversity of values that human cultures have devised over millennia. That is why Moscovici (1978: 55) sees the ideological distinctiveness of ecologism in its commitment to the quality of life rather than to the quantity of material production. Ecologists contend that market economies have failed to value not only nonhuman habitat but also human communal solidarity and social justice. They try to integrate critiques of alienating work conditions with proposals for conserving resources. They charge that representative institutions and bureaucracies are overcommitted to economic growth and undercommitted to goods such as beautiful landscapes and biological diversity. They see local environmental degradation in the context of concerns for world peace. They promote participatory practices not only to mobilize people on behalf of ecological issues but also because they prize a human personality that is active, socially aware, and attentive to the needs of others. The normative quest of French ecologists is ethically pluralistic. It leads to no metric, no prioritizing scheme, that would reconcile all values. Doubt about the timeless essence of either the human subject or "nature" makes it possible to talk of "humanizing nature" without implying an anthropocentric project of seeing "nature" simply as raw material to satisfy human desires.

This multivalent awareness of "the natural question" has been made possible, Moscovici argues, by the developments in nineteenth and twentieth-century science that led him to characterize our state of nature as "cybernetic." Neither an organic nor a mechanical worldview supports the general characteristics of Moscovici's own theorizing: an assertion of the substantive unity of nature and society, attention to the mutability of matter, and a call for "governing" nature. As the next chapter will emphasize, the nature envisioned in a systems approach does all of these things.

3

Systematizing Nature

France's ecological movement began, says Marc Abélès (1993: 9), as a form of "ruralizing romanticism." But it became a movement of lasting political significance only when it incorporated a scientific discourse showing that "the planet is a complex system of interactions" (ibid.: 10). So influential has the systems view been among French ecologists that one group of French sociologists call it "the spontaneous philosophy of post-modern ecology" (Alphandéry et al. 1991: 117). The systems approach, they explain, "starts from a desire to go beyond the traditional scientific division between the exact sciences and the human sciences, in order to apprehend the complexity of relations between nature and culture. . . . It wants to supply, through intersecting concepts such as information and energy, order and disorder, complexity and self-organization, feedback, regulation and entropy, a total vision that grasps the unity of the fundamental mechanisms of nature, man and society." (ibid.: 116) None of these concepts are, in themselves, unique to French ecologism. At least since 1953, when Eugene Odum's *Fundamentals of Ecology* was published, the systems approach has informed the international research program of scientific ecology. To the extent that that research program has yielded knowledge about unsustainable levels of resource consumption, the adverse of effects of monocultural crop production, and human-induced climate change, it is only to be expected that political ecologists around the world would draw upon it. Certainly concepts borrowed from systems theory are prominent in the green thought of English-speaking countries.

Two factors change, however, as one moves into the world of French ecologism.

First, one notices that in France ecologists of almost every stripe take on the language of the systems approach at some point (Prendiville 1993: 130; Simmonet 1982: 13; Jacob 1995: 43–56). Among English-speaking ecologists, systems ecology usually competes for influence with ideas drawn from older environmental traditions. Henry David Thoreau and John Muir, for example, introduced Americans to a sort of respect for nature grounded not in the science of ecology but in immediate experiences of untamed landscapes. Nonanthropocentrists and anthropocentrists alike appeal to this tradition when they connect environmental values to the direct perception of the wild. In Britain, environmentalist ideas sometimes emerge out of cultural preservationists' efforts to protect the "English" characteristics of the pre-industrial countryside (Macnaghten and Urry 1998: 35–38; Barry 1999a: 99–104). In that case, one hears more of the heritage of hedgerows than of feedback and entropy. The pervasiveness of systems theory in France, however, means that theorists more often develop their perspectives by amending its concepts, rather than those advanced by wilderness enthusiasts or pastoral environmentalists.

The second difference concerns how the systems approach is taken up theoretically in France's humanistically charged rhetorical field. In the English-speaking world, systems ecology has regularly been swept into debates for and against nature's intrinsic value. The drive to center ecologism sometimes hides a fateful ambiguity in the systems perspective: Its holistic view of nature potentially validates both nonanthropocentric and anthropocentric extremes. Nonanthropocentrically, humanity can be fully naturalized as part of an ecosystemic whole, i.e., seen as just one more species within nature and subject to its laws. Anthropocentrically, nature can be more fully humanized than ever before. By virtue of understanding ecosystemic relations, humanity learns how to subject nature to its purposes with impunity. Mindful of this ambiguity, French ecologists attempt to take advantage of the insights of scientific systems ecology while consciously resisting its potential to spawn a centered ecologism. They do so with varying degrees of consistency and success.

René Passet keeps his use of systems theory off center by emphasizing the irreducible diversity of environmental values. This diversity confounds any attempt to derive values from nature or to reduce environmental value to economic value. Joël de Rosnay tries to follow a similar path but stumbles into the ambiguity of the systems approach. Edgar Morin offers the most sophisticated use of a noncentered systems approach. Determined to mitigate the potentially authoritarian political implications of systems ecology, Morin employs psychology, phenomenology, and a quasi-Hegelian historicism to deny closure to any potential center of green value.

Systems and Values

How botanical studies, "scientific natural history," and population studies merged with information theory, cybernetics, and mathematical modeling of physical systems and how Ludwig Von Bertalanffy drew these strands together into a general systems theory are stories that go beyond the limits of the current study. What is important here is recognizing that the systems approach has often driven the research program of scientific ecology (Kitching 1983: 1–5). Mathematical modeling based on systems concepts has made it possible to understand fluctuations in the populations of various species and to relate those changes to the organisms' reproductive behavior and to environmental conditions, has shown how phenomena such as human-caused river pollution have created positive feedback loops that deplete aquatic life in estuaries, and has supported observations that feeding a human population of more than 10 billion in the twenty-first century may require simplifying ecosystems on large parts of the planet's surface, thereby exacerbating vulnerabilities to the catastrophic spread of plant and animal diseases, to pollution from pesticides and fertilizers, and to energy shortages.

The following summary of the major tenets of a systems approach sets the stage for a discussion of particularly French ways of using it.

(1) Phenomena should be understood holistically, as goal-oriented totalities, not just analytically, as effects of causally linked elements. "In contrast to the analytical method," explains Dominique Simmonet (1982: 13–14), the systems approach "does not isolate the subject of study, but seeks to consider the interactions in the system where it is found. The system is conceived as a complex, organized entity composed of elements and relations." Analytical approaches study phenomena by breaking them down into simpler, better-defined component parts which are related in linear chains of cause and effect. The systems approach "considers a system in its *totality*,

its *complexity*, and according to its own *dynamics*" (Rosnay 1975: 119). What matters to the character of a system is not the properties of its isolated components but its structure: how its components functionally support one another to form a whole. By virtue of the complexity of interactions among its parts, a system develops properties (e.g., the ability to be self-regulating) that cannot be explained solely by the properties of their building blocks. Thus, the system as a whole appears goal-directed. In pursuit of that goal, relationships between the system's parts can be flexible and irregular.

(2) As the science of self-regulating systems, cybernetics shows how systemic properties are stabilized—or disrupted—by flows of matter and energy interpreted as *information*. That is, cybernetics studies how systems, including life forms, respond to their environment not merely by reacting mechanically to forces impinging on them but also (and more important) by processing inputs as messages. Matter and energy entering the system ("inputs") and exiting the system ("outputs") have to be kept in balance in order for the system to persist. "Negative feedback" consists of inputs that provoke corrective, equilibrium-preserving changes in the system. Through negative feedback loops, systems become "self-regulatory"—for example, they tend to preserve a certain temperature, a certain concentration of chemicals, or a certain level of population. Positive feedback consists of inputs that continually reinforce a single direction of change, thus throwing a system out of equilibrium and toward growth or death. Open systems exchange energy and material with their external environment (Robin 1975: 75).

(3) Complex systems are characterized by *dynamic* stability. Systems preserve their identity through a continual process of self-renewal. Unlike a crystal, which remains self-identical in the sense that its constituent molecules remain the same and are locked into certain organizational patterns, an organism keeps its identity by "maintaining its structure and its functions through a multiplicity of dynamic equilibria" (Rosnay 1975: 129). Likewise, entire ecosystems consist of dynamic equilibria between numerous species of living organisms and cycles in the nonbiotic environment. Complex systems are *homeostatic*: they "react to environmental change . . . by a series of modifications that are equal in size and in the opposite direction to those that gave rise to them: the goal of these modifications is to maintain the internal equilibrium" of the system (ibid.: 129). In this way,

living systems acquire a degree of autonomy from their environment while remaining in constant interaction with it. Within certain limits (beneath certain "thresholds"), they are able to maintain their own structure and functions, even in the face of environmental change. This has an important implication for ecological thinking, expressed in Henri Atlan's rule (1972: 54–55): "For a given value of the variety of possible disturbances, if one wants to bring a system back to a limited number of states, the number of available regulative responses must be larger as the number of acceptable different states is limited." Homeostasis in living systems depends on complexity. Adaptive success—from cell repair to searching out new food supplies—requires energy. An absence of available energy makes adaptation impossible. Therefore, systems theorists pay particular attention to the energy flows that power homeostasis.

(4) All exchanges between systems and their environment are subject to the laws of thermodynamics. The second law of thermodynamics holds that energy has an irreversible tendency to dissipate. Physical processes thus move from order to disorder. "Entropy"—a measure of disorder—is, in the long term, unavoidable in isolated systems. Biological entities—ecosystems as well as organisms—are "negentropic" (Robin 1975: 57; Lebreton 1977: 51). Open systems capture energy from their environment and use it to resist entropy. Their immense variety of elements, grouped hierarchically in different levels with different functions, allows them to stabilize their form, to adapt, and to evolve. This is as true of humankind's increasingly complex forms of social organization, which have required harnessing ever larger quantities of energy, as it is of rainforest ecosystems (Georgescu-Roegan 1971). An inevitable consequence of negentropic activity, however, is the increase of entropy in the system's surroundings. This entropy is in the form of waste products and unusable energy (Daly 1992). "In the final analysis," writes François Jacob (1970: 272–273), "the maintenance of a living system in a given state has to be paid for; a return to equilibrium, which is always unstable, is paid off by a deficit of organization in the surroundings, that is, an increase of the disorder in the whole constituted by the organism and its environment." In the end, entropy wins: the organism dies. Humanity's ability to stave off its own entropic demise depends on harnessing and managing energy in ways that do not ultimately destabilize this planet's life-sustaining ecosystems.

From studies of homeostasis to thermodynamics, a systems approach seems to overflow with lessons for political ecology. It explains how seemingly "spiritual" properties such as goal-directedness result from complexity itself, thus revising the mechanistic world view of modern science. It suggests that mankind's transformations of the environment (e.g., reducing the number of plant and animal species, draining wetlands, generating nonbiodegradable waste) threaten to destabilize life-sustaining ecosystems by reducing their complexity. It demonstrates the need to draw on energy to develop complexity at the same time that it traces the adverse global effects of our rapid exploitation of energy-rich resources.

Yet there is no single route leading from systems ecology to an ecological political theory. No matter the complexity or fragility of the systems in which human activities are embedded, judgments must be made about the *value* of the phenomena discussed. English-speaking ecologists often press to understand the ground of such judgments: Whose interests or well-being are served—our own or nature's—when we act to preserve complex ecosystems? When systems ecology enters the rhetorical field of English-speaking ecologism, it is made to serve a centered ecology. In fact, it serves *both* centers.

The highly influential 1972 report *The Limits to Growth* offered a straightforwardly anthropocentric application. Using models derived from systems theory, its authors projected that unconstrained economic growth would run up against insurmountable obstacles in the long term, owing to the finitude of Earth's nonrenewable resources and the planet's limited ability to absorb pollutants. "Positive feedback loops" of human population growth and the growth of industrial capital could surpass the planet's carrying capacity, leading to "a sudden and uncontrollable decline" in population and industrial capacity" (Meadows et al. 1972: 23). This report referred to matters of prudential human concern and made no reference to "nature's intrinsic value." Its ethical assumption was that human well-being had value and that such value was endangered by environmental problems.

Eugene Odum (1974) took a step in the direction of asserting nature's intrinsic value when he suggested that Aldo Leopold's "land ethic" fits perfectly into a cybernetic analysis of our environmental predicament. If human population growth, technology, and economic development are putting increasing stress on man's natural life-support system, and yet if those stresses are not themselves providing enough "negative feedback" to make us reduce environmental damage, then what is needed is a profound shift in our ethics. "The extension of ethics to include man-in-environment relationships must become an integral part of man's philosophy," said Odum (1974: 14), "if for no other reason that such an extension provides the needed internal governor." Still, Odum's argument was more utilitarian than decisively nonanthropocentric. He contended not that nature *has* intrinsic value but rather that it would be *useful* to humanity if we were to act as if it did.

The case of James Lovelock illustrates the next step on a path toward ecological centering. The author of the much-discussed "Gaia hypothesis" started from the same anthropocentric point as Odum. Gaia is the oceans, the atmosphere, rocks and soil, and all the life forms on Earth in dynamic interaction. According to Lovelock's hypothesis, together Gaia's components may constitute a "superorganism" whose parts interact homeostatically to maintain conditions conducive to life. In 1979, Lovelock argued that better scientific understanding of Earth's ecology would allow us to identify and protect Gaia's "vital organs." Some of Gaia's subsystems, such as the continental shelves or the tropical forests, may play such crucial roles in circulating and transforming substances on which life depends that disrupting their functioning could threaten Earth's ability to sustain life— especially human life. Lovelock advised setting such regions aside from human exploitation until they were better understood. His admonitions to avoid destroying Gaia's vital organs were rooted in a concern for human health and material well-being. In 1995, Lovelock added a new preface to his work in which he admitted that prudential reasoning no longer seemed adequate. Now claiming to have dropped his "loyalty to the humanist Christian belief in the good of mankind as the only thing that mattered," he argued that we are "part of a community of living things" and that we have "only obligations" toward that community (Lovelock 1995: viii).

English-speaking environmental ethicists of a more philosophical bent have gone much further in developing this view. Nonanthropocentrists, we have already seen, maintain that human values must be seen as internal to the values of a larger, ecosystemic whole. Freya Mathews (1991) makes such an argument ontological. The idea of intrinsic value, she maintains, arises out of the very logic of theories of "self-realizing systems." Self-realizing systems are organisms, or complexly interrelated sets of organisms, that seek to maintain a steady state with respect to external variables. They do so by meeting their own energy requirements, repairing their tissues, reproducing, and so on. Every such system strives for self-maintenance and therefore can be said to *value* its own self-perpetuation. "With the appearance of systems-for-self-realization," reasons Mathews (ibid.: 103), an "ontologically new quality—the quality of being-for-itself—enters the world." The value-creating status of being-for-itself does not, pace Jean-Paul Sartre, enter the world through an intending human consciousness; value can be defined systems-theoretically and without reference to humanity. In other words, systems theory *is* a theory of nature's intrinsic value because value, intrinsically, is a property of self-realizing systems.

What is striking in these uses of the systems approach in English-speaking ecologism is how each leaves the nature/humanity dichotomy outside the scope of reflection. Meadows and the early Lovelock assume that there is (on the one side) an objectively observable set of natural processes and (on the other side) human beings who evaluate the impact of those processes on their well-being. They picture us as knowers who examine nonhuman systems from the outside and make decisions about their value as a function of human interests. Nonanthropocentrists such as Mathews purport to go beyond anthropocentrism by discovering value-conferring properties in natural processes themselves. Like the anthropocentrists, however, they suppose that a systems view allows them to rise above ecosystems and (ideally, at least) to see them whole. It is from that holistic understanding that they claim to infer the well-being needs of nonhuman entities.

A pragmatic anthropocentrist (in the ranks of English-speaking ecologists, one of these is never far away) stands ready to moderate such conclusions. Bryan Norton finds the nonanthropocentric interpretation unconvincing. He defends a "contextualist approach to environmental management" in its stead (Norton 1991: 148, 154). More like a noncentered theorist, a contextualist such as Norton is comfortable accepting "science as a value-laden enterprise" (ibid.: 193). But this concession does less to mix humanity and nature than is apparent at first sight. It turns out that "systematicity" and "dynamism" are among the "axioms" that constitute "the scientific portion of a new and emerging worldview" (ibid.). The truly evaluative work gets done, not when scientists formulate such axioms, but when environmental managers apply them to protect "ecosystem health." Nature, revealed scientifically as a set of self-identical processes, remains the focal point of human valuation.

What might be missing when debate proceeds in this way? Examining the works of French ecologists who also use systems theory might make us aware of several suppressed or deferred lines of inquiry. In France, the ambiguity of the nature/humanity distinction is woven into systems-theoretic reflections. Theorists examine how the exercise of power in human communities is bound up with the very terms used to describe "nature." They probe the psyche of the human subject who will make decisions about either respecting nature or managing it. Rather than maintain that through a systems approach we have finally understood our place in nature (because now there is a place in nature for complexity, goal-oriented adaptation, autonomy), or that a systems approach furnishes a new view of nature from which we can deduce moral obligations, French ecologists call attention to how human values interact with scientific understandings of the natural world.

Being and Having: Economics in a Systems-Ecological Perspective

The systems approach took root in France largely as a result of the efforts of the "Group of Ten." That group's predecessor—an intellectual club called Objectif 72, created in 1965—sought to bring awareness of scientific and technological advances into political discourse. Jacques Robin, Robert Buron, Henri Laborit, and Edgar Morin (each of whom had attended some meetings of Objectif 72) were the core of a larger group of economists, computer scientists, and sociologists who, beginning in 1967, assembled regularly to converse about relationships between technology and society. The starting point of the Group of Ten's discussions was this question: "If we took account of the new knowledge in the fields of computer science, biology, and neuropsychology, would this help decision makers to some degree to rid political discourse of its magical elements?" (Robin 1990: 82) Like Enlightenment thinkers, the Group of Ten hoped that human intelligence would be able to penetrate the mysteries of the surrounding world and use its new-found knowledge to drive out superstition. Giving their project a political tinge from the very start, the Group of Ten anticipated criticizing democratic political leaders and the institutions that shape their discourse.

In the 1970s, participants in the Group of Ten published a variety of works elaborating on the systems approach and applying it to environmental political issues. Henri Atlan (1972) advanced information theory with the idea of "organizing noise" to explain how information-conveying signals arise out of physical disorder. Henri Laborit wrote about the levels of organization in living things, memory circuits, and the constructive place of the nonrational in human societies. His notion of "signifying information" applied to both living and nonliving matter—including society and culture (Laborit 1968, 1974). Approaching a more political level of observation, Laborit turned to criticizing individualism, the dominant ideology of an industrial society given over to the pursuit of material abundance. Jacques Robin (1975) used a systems approach to the environment to challenge conventional measures of economic growth. Expanding on that critique, René Passet's 1979 book *L'Économique et le vivant* [*Economics and the Living World*] became one of the most influential works of French ecologism using a systems approach.[1]

Passet (1979) contrasts the different logics of two types of holistic entities: biotic systems and human economic systems of production and exchange. Studies of economic systems view people as autonomous, interest-maximizing individuals, whose worldview is governed by an ethic of "having" (possession). Millions of such individuals, linked in exchange, constitute markets that allow them to signal their preferences to one another, to arrange trade efficiently, to innovate, and to impose order on nature. But economists' conceptions of market equilibria, Passet insists, ignore the characteristics of biotic systems. Economic analysis takes no account of the regulatory mechanisms—stabilizing reserves of minerals, energy flows, sources of feedback, complex layerings of diverse organisms—that make it possible for living systems to reproduce themselves. Linear models of production and exchange omit such nonlinear features of environmental dynamics as synergistic interactions of pollutants and thresholds for species reproduction. Limiting itself to "the domains of production, exchange and consumption of 'useful and scarce' things," economics is condemned "to develop a logic and to motivate actions that set themselves up in formal contradiction with those of the biosphere" (ibid.: 17).

Economic relations, Passet concludes, should be reconceived so that they are seen as a subsystem within the larger spheres of human activity and the biosphere. In that larger context, human finality is "being," not having. "Being" includes not only physical survival and reproduction but also existing socially—internalizing a wide-range of group norms (ibid.: 111–114). Norms of justice, respect for life, beauty, emotional expressiveness, and deference to the sacred rightly circumscribe individualistic choices made in the marketplace, because such norms give richness and meaning to collective life. Politics—by which Passet means the legislative activity through which such norms are given form for society as a whole—thus takes precedence over economics. "Political power, open to all citizens, remains indispensable at all levels of the Nation, to define the content of a social project which is never reducible to the imperatives of the productive sphere alone." (ibid.: 227) In practice, Passet's politics would entail using a systems approach to understand the conditions of reproduction of the ecosphere. Scientific ecologists could identify the "quantitative norms in whose limits the problem of the rational management of resources is located" (ibid.: 227). Legislation would then be devised to restrict economic activity in light of these norms. But Passet's concern for "being" indicates that such legislation would somehow have to be shaped so as to take account not only of systemic limits but also of the community's other aesthetic, religious, and political-ethical values.

In one of his closing paragraphs, Passet alludes to the humanism of this perspective: "The real questions concerning Man—obscured for centuries by the struggle for survival—are finally being raised for him: they refer to his being, his place in relation to the order of things and his relations with Nature." (ibid.: 240) Doesn't that last phrase simply perpetuate the nature/human culture divide in its most problematic form? If we have "relations with Nature," isn't Nature distinct from us? Furthermore, doesn't a humanistic focus on "*human* finality" posit human pre-eminence over the nonhuman world? And doesn't that assertion of pre-eminence lead Passet to defend what nonanthropocentrists often deplore: a policy of "the rational management of resources" that potentially subordinates everything in our surroundings to human control? Perhaps so, since Passet undoubtedly has substantial confidence in the ability of ecological scientists, as observers of "Nature," to discover the reproductive requirements of living systems.

But Passet's references to "being" suggest a different conclusion. Passet's humanism expresses a deliberate effort to place the problem of what nature is, and of what human beings are, in the broadest possible context. In that context, an ecologist's understanding of the reproductive limits of a "natural" system cannot be dispositive with respect to social policy, since science "cannot claim to demonstrate the . . . superiority of a vision which . . . engages man in all his dimensions" (ibid.: 240). Being alludes to a pluralism of values through which we determine what aspects of our world are worth preserving. A systems view does not simply enclose the domain of human choice, like a fence protecting us from wandering unwittingly into a dangerous bog. The role of science in human decision making cannot be that simple, because, as Passet notes, "by transforming the world man transforms himself" (ibid.: 106). Because our behavior is affected by the qualities of our surroundings, which we ourselves have altered, it is misleading to theorize in a way that creates separate categories of "nature" and "humanity."

What is critical is that our understandings of the systemic relations of living things become part of a multivalent, deliberative assessment of our "social projects." We do not reach such an assessment by deducing policy either from scientific data or from a set of principles, whether economic or juridical. The very complexity of Passet's humanistic value pluralism means that he neither accords "nature" intrinsic value nor interprets its value purely as value-for-humanity.[2] His receptiveness to a wide range of values, combined with a determination not to rank them or give them ultimate grounding, gives him affinities with American environmental pragmatism— a variety of English-speaking ecologism that, unusually, claims to be neither anthropocentric nor ecocentric (Weston 1996: 285).

Yet not even pragmatists always escape the influence of a rhetorical field of centered ecologisms. A comparison of Passet's environmentalist critique of economic reasoning with Mark Sagoff's 1988 book *The Economy of the Earth* shows how differences in the ecological rhetorics of the two linguistic communities eventually show through in decisive ways. Like Passet, Sagoff is at pains to insist that environmental issues should not be addressed exclusively from a perspective that measures success in terms of efficiency. Like Passet, Sagoff eschews the individualism of a worldview that looks only to satisfying people's preferences. And, like Passet, Sagoff defends the appropriateness of having deliberative, political processes set the norms within which economic transactions take place.

But Sagoff feels it necessary to push his argument one step further so that he can account for the special significance of wilderness and for assertions of nature's intrinsic value. For Sagoff, it is not sufficient to insist that a wide range of values, as well as the best current knowledge from scientific ecology, should inform environmental debate. It is necessary for the philosopher to anticipate the outcome of such debate by explaining what makes certain things intrinsically valuable. Discussions about preserving species or pristine territories are properly framed, he contends, in terms of the *cultural meaning* of those things. He notes that eagles, untamed rivers, and virgin forests express metaphorical qualities—freedom and integrity—that are part of "our" character as a people (Sagoff 1988: 128). Historians of culture can "discover" these noneconomic meanings in a nation's literature, religion, and art. Sagoff concludes that citizens' interests in protecting national character—not only consumers' willingness to pay for certain goods—are at stake in environmental policy. He supports an environmental movement that "asserts the importance of . . . cultural, historical, aesthetic, and religious values," not one oriented only toward "the 'satisfactions' or 'benefits' it offers human beings" (ibid.: 148–149). As Sagoff embroiders this seemingly culturalist, human-centered argument, strands of intrinsic value reasoning repeatedly get woven in. He suggests that writers do not merely impute virtues to nature; they discover them there (ibid.: 134, 141). Moreover, he associates his environmentalism with Aldo Leopold's view that "nature . . . itself has a moral worth and therefore should be protected for its own sake" (ibid.: 149). Which is it: Are nature's values in nature, *there* for us to discover, or are they in us, in the symbols that constitute our collective identity?

Sagoff's combination of culturalist reasoning and the assertion of nature's intrinsic value virtually invites charges of theoretical self-contradiction. Anyone searching for the center of environmental value will find pragmatic, culturalist reasoning unsatisfactory (Katz 1997: 67). We need only reflect that, whereas bald eagles may plausibly be said to represent Americans' values of freedom and vigilance, spotted owls have had no such long-standing significance in the American popular imagination. The implication would seem to be that an Endangered Species Act should mandate saving bald eagles but not necessarily spotted owls. The variability in the ways that Sagoff's environmentalism might treat different species reveals that he is defending noninstrumental values in *American* culture, not values intrinsic to nonhuman entities. In a linguistic community where "arguments have come increasingly to be cast in terms of the interests of nature itself" (Goodin 1992: 8), it seems, theorists sometimes find it difficult to resist the polarizing force of its rhetorical field—even if it causes them to skip from one ecological center to the other.

Another troubling aspect of a culturalist argument is that it tends to displace entirely the role of scientific ecology in defining "nature." Sagoff does not deny the relevance of systems ecology to decision making in environmental affairs; he merely rejects claims that its methods constitute the sole path of "rational" inquiry. As a pragmatist and a democrat, he holds that inquiry can be rational so long as it "emphasizes virtues of clarity and open-mindedness" (Sagoff 1988: 12–13). This conception of rationality helps to legitimize forms of environmental policy making that include deliberations by ordinary citizens, not just experts. The problem is that a culturalist interpretation of intrinsic value risks making environmental policy simply a reflection of social identity. We have to ask if a largely aesthetic-historical approach to environmental issues does not divert our attention from the complex interconnections that systems ecologists wish to call to our attention. Culturally, wetlands may be little more than dismal swamps. Ecosystemically, however, they are crucial links in the circuit of life: they are spawning grounds for numerous species of plant and animal life; they filter out waterborne pollutants; they resupply underground aquifers; they mitigate floods. Ignorance of these systemic functions, which is often culturally encoded, contributes to environmentally destructive strategies of economic development. For purposes of deciding what most deserves protection, Sagoff's environmentalism is in danger of viewing "nature" only through a cultural lens.

From Passet's viewpoint, nature is not so subjective. Ecological thinkers have to make room for a scientific-realist moment as they try to evaluate the processes through which humanity transforms the world and, reciprocally, is itself transformed. He proposes no unifying theory of value, culturalist or otherwise, through which to process scientific information. His appeal to "being" signals a determination to use the scientific realism of the systems approach while holding at bay any tendency to let its authority override multivalent human judgments. On that difficult path, Passet keeps his balance reasonably well. He owes that balance, we shall see, to his awareness that science and politics compete as forms of power. Not every systems theorist, French or English-speaking, does so. The writings of Joël de Rosnay illustrate the dangers of a politically less cautious use of a systems approach.

Symbiosis and Technologized Nature

Joël de Rosnay was at one time a researcher in biology and computer science at the Massachusetts Institute of Technology. Later he was Director of Research Applications at the Pasteur Institute. Today he is a director of the City of Science and Industry near Paris. He was a relative latecomer to the Group of Ten in the early 1970s. But through works such as *Le macroscope* (1975) and *L'homme symbiotique* (1995), Rosnay has become probably the most widely read advocate of ecological systems theory in France.

For Rosnay (1975: 22), the interrelations that make up an ecosystem mean that "in a certain way, [an ecosystem] is a living organism." Like an individual organism, an ecosystem captures energy, circulates nutrients, and spontaneously adopts stabilizing strategies in response to perturbations. Rosnay—who explicitly favors Lovelock's Gaia hypothesis—argues that, through the rapid spread of technological developments, humanity has turned itself into a "social macro-organism," or what he calls a "cybionte" (a neologism combining the idea of a cybernetic, self-regulating machine and a living organism) (1995: 143). On roads, railways, and air routes, the cybionte circulates its own vital nutrients. Our machines are like muscles and organs that move and process matter. To register needs or threats, we have developed nerve-like detectors: markets, media, polling firms. Just as sensory impressions are transmitted through the entire nervous system of a living organism, we have covered the globe with information networks. Coordination of all of these subsystems is in part deliberate (e.g., the regulatory activities of governments and international organizations) and in part spontaneous (e.g., the decentralized "parallel processing" of information accomplished by individuals and firms attempting to adapt to local changes). All these components are linked systemically. Our interdependencies give us the potential to become organized in such a way that we respond adaptively to breakdowns in any particular subsystem in order to preserve the cybionte as a whole. According to Rosnay (1995: 108), "the general theory of . . . the dynamics of complex systems . . . shows that people can contribute to the synthesis and emergence of a planetary macro-life that exists symbiotically with the natural cycles of the planet." Symbiosis denotes a co-evolutionary relationship of mutual advantage.

But we have not yet realized this symbiotic potential: "Gaia is sick. Man acts like a parasite endangering its equilibria. The greenhouse effect is Gaia's fever. The hole in the ozone layer is the cancer of the 'skin' that surrounds it. Acid rain, solid waste and water pollution are its digestive illnesses." (Rosnay 1995: 153; 1991: 65–66) On environmental issues, Rosnay has sought to alert a wide public to the folly of unbridled exploitation of the biosphere. As early as 1975, he criticized the preferred modern strategy for increasing the complexity of our social systems: not by more efficiently using energy available in the biosphere, but by drawing on energy stockpiles—fossil fuels—that have long been environmentally sequestered. In the long term, he warned, this strategy for social adaptation must fail. Energy stockpiles will be depleted, and throwing the waste products of combustion into the biosphere may introduce disorder into atmospheric and ecosystemic equilibria around the globe.

What would be the ecological politics of "symbiotic man"? Might the systems view predispose theorists to favor centralized planning? After all, isn't there a certain logic in claiming that if humankind spontaneously over-exploits its environment it needs a "governor" to supply the regulatory mechanisms that it lacks by nature? Although many critics see an authoritarian bent in systems-theoretic ecology, Rosnay prefers to play up its potential to support political decentralization and democracy (1975: 214–224; 1995: 205–209). Through computers and communications networks, citizens can now more easily become informed about problems and provide "feedback" on them. They can also act like computers linked through parallel processing—solving multiple problems and generating intelligence superior to that of any single component. With more accurate information coming from below, it should be possible to adjust energy consumption, recycling, and related policies so as to remain within the carrying capacity of the local environment. At the same time, "macroscopic" political representation is still desirable, according to Rosnay. Representatives perform functions of "filtering" and negotiation that are needed to avoid socially destructive, self-amplifying oscillations that would probably occur if public opinion were directly translated into policy.

Ethically, Rosnay calls for individuals to develop a sense of community appropriate to their place in the symbiotic unity of Gaia and the cybionte: "New values and an ecological ethic are necessary in order to cross the barrier of individualism and to open out onto an organized collectivity that respects individual liberty and personal initiative." (1995: 162) He complains that modern economies, with their emphasis on unlimited growth, international competition, and consumer satisfaction, are virtually machines for manufacturing egoists. Rosnay would promote an altruistic "new humanism" (ibid.: 323), an ethic that values creativity and social solidarity.

Programmatically, Rosnay (1975: 132–139) presents humanity with the "ten commandments of the systems approach." These include preserving variety (because simplifying ecosystems destabilizes them); protecting especially sensitive points of homeostatic systems; and allowing some forms of environmental "aggression," such as forest fires (since this permits ecosystems to evolve). In 1975, Rosnay imagined that such policies might be realized in a steady-state "ecosocialist" society. In the more conservative 1990s, he spoke of "regulated, adaptive development" (Rosnay 1995: 200) rather than "sustainable development." Thanks to information networks and the cooperative "ecoethics" of citizens, regulated, adaptive development better expresses ideas of ongoing technological change and of economic "self-regulation." Still, substantial traces of ecosocialism persist, even if the label does not. Taken seriously, Rosnay's calls for combating unemployment with "nonmarket activities" (including "mutual education, social assistance . . . and voluntary efforts to build up eco-capital") and his description of a "symbiotic economy" to "redistribute resources, reduce inequalities and reinvent work" on an international scale would require massive political changes of a sort often associated with ecosocialism.

Rosnay's attempt to think through the conception of humanity and nature needed to support these changes reveals a novel, albeit ambiguous, dimension of his ecologism. For Rosnay, the relationship between the cybionte and Gaia creates "a superior totality abolishing the traditional borders between natural and artificial" (1995: 72; cf. 22, 92, 284). In his eyes, it is not that the traditional borders were always unclear or indefensible. Rosnay obviously has considerable—perhaps inordinate—confidence in the ability of the sciences to understand natural processes. But advances in that understanding, he argues, are now leading to so many border crossings—from genetically engineered bacteria to species redistribution across the globe to, someday, direct connections between human brains and computers—that the very notion of a frontier between the natural and the artificial is becoming meaningless.

Implicit in this claim is a human history of nature. Rosnay is saying that the idea of "nature" long reflected humankind's relative technological impotence. As our technological prowess increases, "nature"—for better or for worse—loses its conceptual power to express the idea of "all that is nonhuman." Rosnay replaces the traditional dichotomy of nature and humanity with a systems view of the two. A systems approach, in this view, is not a new understanding *of* nature; it is a unifying mode of thought that enables us to grasp relationships both within and between what were formerly called nature and society. And, although Rosnay believes that systems theory helps us better appreciate the complexity, significance, and fragility of ecosystems, it does not lead him to declare that any part of nature deserves respect because it embodies an "end in itself." Systems theory decenters our way of conceiving ourselves, preparatory to designing new ways of integrating a wide range of human interests and environmental constraints. In practice, Rosnay probably agrees entirely with the pluralistic perspective of René Passet (whose work he often cites). Thus, there is in Rosnay's work a glimmering of what is characteristic of French ecologism: a sense of a "humanized" nature that leads toward an ecologism dedicated to rethinking the very frameworks of thought that have forced us to fit ourselves into a dichotomy of natural and human things. But only a glimmering. Repeatedly, Rosnay reverts to the traditional dualism—and in ways that raise disturbing political-ethical questions about his work. When he says, for example, that systems theory offers "a new vision of nature" (1995: 34), he continues to allude to what nature really *is*. In the process, he replaces an old dualism with an old monism. He naturalizes humanity. He submerges it in the symbiotic couple of the cybionte and Gaia. In that case, a "humanist morality" really consists of the values that we *have to* adopt in order to maintain the vital functions of this "planetary superorganism" (ibid.: 147). "A humanist morality," says Rosnay, "can be born from the necessity of assuring human liberty and the exercise of responsibility in the framework of natural laws progressively brought to light by science" (ibid.: 322). On one hand, then, nature limits our liberty.

Yet Rosnay's account of the world-transforming activities of modern science makes the very idea of "natural laws" extremely problematic. If humankind is "abolishing the traditional borders between natural and artificial," how can Rosnay even speak of "natural laws" as a "framework" to contain human action? What would be an example of Gaia's laying down the law? Rosnay brims over with technological optimism, so that virtually any ecological limit seems surmountable (at least for human, practical purposes). Is too little food available to feed people in certain parts of the world? By 2058, Rosnay imagines, we may use miniature "biorobots" to harvest genetically engineered foodstuffs from even the most difficult terrain (ibid.: 276). Are we worried about our dependence on limited supplies of fossil fuels and about the pollution associated with them? A planet-wide transition to solar power and the use of hydrogen as a motor fuel will put those concerns to rest (ibid.: 268 ff.).

Although Rosnay opposes nuclear power and worries briefly about the spread of "electronic drugs" based on biofeedback techniques, he is generally far less cautious than others in the Group of Ten. For instance, Jacques Robin worries about the genetic manipulation of plants and animals (1989b) and charges that the global economic system encourages "hyper-competition," which leads to "the abuse of techniques, their use without measure," causing "ecological disturbances world-wide" (1989a). Rosnay's ecological perspective simply does not include many considerations that, for others, argue against some technological remedies (a desire to preserve certain territories in pristine condition; a suspicion that complex technologies routinely produce environmentally undesirable side effects; a belief that human communities are best served by simpler, more "convivial" technologies). On the other hand, Rosnay's humanism builds upon the other half of the traditional dualism: the idea that a free, knowing consciousness makes choices as a function of its own purposes and sense of moral responsibility. Talking of "human liberty and the exercise of responsibility in the framework of natural laws" implies that knowing "nature" enables us better to bend it to human purposes. Rosnay's imagined technological remedies to problems of food scarcity and climate change seem to suppose a world in which nothing is, in principle, preserved from human alteration. But what *are* our purposes? Even though symbiotic man is free, Rosnay fails to ask critical questions that are as old as philosophy itself. What makes life meaningful? What makes life (in the cybionte) worth living? Before we applaud "the progressive movement of the human species toward organizational levels of higher complexity" (Rosnay 1995: 183), we should join René Passet and many others in asking about the effects of ever-increasing organizational complexity on our sense of personal fulfillment, on our ability to draw inspiration from our surrounding world, and on our sense of involvement in decisions concerning things we care about most. Scarcely recognizing that such questions exist, Rosnay truncates the philosophical significance of his work.

Rosnay's metaphysical ambivalence helps explain the tacitly technocratic orientation of his politics, for free consciousnesses who perceive the "framework of natural laws" are only a segment of humanity. If it is scientists and engineers who understand the system-destabilizing tendencies caused by our species' technological prowess, then it is hard not to conclude that they are best placed to act as "regulators." In Rosnay's own language, they "pilot" the social system toward environmentally sustainable practices by favoring new technologies (e.g., genetically engineered plant and animal species) and by administering the proper "dosage of constraints" (1975: 136).

It is understandable, then, that Rosnay has been criticized for proposing a potentially authoritarian version of ecologism. Detractors claim that Rosnay leans toward a "green Leviathan" that pays little attention to questions of the equitable distribution of resources (Paraire 1992: 213–214; Jacob 1995: 55–56). Others have cautioned that systems theory opens up an enormous gap between ecological activists who see their objectives as socially liberating and those who understand that the ecological project is essentially regulatory (Faivret et al. 1980: 59).

Rosnay would reject such accusations. He praises liberty, "self-regulation," and democracy. His optimism about the benefits of technological adaptation and about uncoerced human cooperation allows him to skirt the authoritarian conclusions of certain kinds of neo-Malthusian reasoning. Still, Rosnay's words do not settle the issue. His optimism can be dismissed as naive, and his easy talk of "man" becoming the "co-pilot" of the living world obscures questions about who will actually be steering the system.

What is missing from Rosnay's use of systems theory is a set of political reflections that, in effect, knock ecological thinking off center. Those reflections appear in the works of Rosnay's more circumspect colleagues in the Group of Ten. René Passet (1979: 230) warns that the systems approach inherently disposes its practitioners to think of individuals as replaceable—sacrificeable—units in a larger whole. That is why he not only praises decentralization (as Rosnay does) but also (unlike Rosnay) puts the whole issue of political deliberation in a framework whose dimensions far exceed those of a systems approach. "To the degree that a value system cannot be demonstrated scientifically," says Passet (ibid.: 237), "there are necessarily many conceptions of social utility. This constitutes, in my opinion, an invitation to pluralism, to tolerance, and a justification for alternating in the democratic exercise of power." In Passet's conception of "being," politics takes precedence not only over economics but also over science, even as knowledge from both of those domains is fed into the community's deliberative processes. Robin (1989b: 202–210) emphasizes that systems become self-regulating only by maintaining a balance between reproducing their own structures and remaining open to external, environmental variations. It is this balance that constitutes a system's autonomy. In regard to human society, this implies that a culture that reproduces itself too efficiently actually hampers its ability to respond to environmental change. From this perspective, Rosnay's cybionte is more alarming than ever. Robin fears that industrial society, with its patterns of audacious technological innovation, mass consumption, and anonymous individuality, has short-circuited the autonomy-renewing potential of democratic organization.

Such political reflexivity can keep systems-inspired thinking from settling on a stable center of environmental value. At one moment, French ecologists understand scientifically the systemic consequences of technologies on the environment; at another, they watch for politically contestable imagery within the presentation of scientific description; on top of all of that, they situate scientific knowledge in a swirl of values whose diversity attests to a context—Passet calls it "being"—that is more encompassing that even the most inclusive ecosystem. Those traits of noncentered theorizing, present in the works considered thus far, receive their epistemological justification through the efforts of the most philosophically ambitious member of the Group of Ten: Edgar Morin.

The Search for "Complex Thought"

Edgar Morin's studies for a university degree were interrupted by World War II. Nonetheless, he managed to make himself into an intellectual in the uniquely French mold. Born in 1921, he became engaged in the defining political movements of his generation. His itinerary took him from the Resistance to membership in the Communist Party to a celebrated departure from the Communist Party in 1951. In 1957, he was co-founder of a nondogmatic Marxist journal, *Arguments*. After its collapse he joined *Socialisme ou Barbarie*, the anarcho-Marxist journal of Claude Lefort and Cornelius Castoriadis. With Lefort and Castoriadis, Morin greeted the May 1968 student and worker rebellion as an expression of socially creative energy. This was also Morin's last foray into revolutionary politics. A research stint in California exposed him to a group of cyberneticians, historians, and biologists debating "the unity of man" (Fages 1980: 203). Bringing his understanding of systems theory back to France, he co-founded the Group of Ten and began to turn his intellectual energy to devising a metatheory of the sciences (Kofman 1996: 7–10). Through all these changes, Morin established critical contact with virtually every major thinker and intellectual trend in postwar France: Jean-Paul Sartre and Maurice Merleau-Ponty, Claude Lévi-Strauss, Jacques Lacan, Louis Althusser, Hegelianism, existential and structural Marxism, and then the Group of Ten's discussions of the systems approach—and ecology.

Morin, an apostle of transdisciplinary thought, is committed to bringing biological, psychological, and cultural explanations of human behavior together in a "fundamental anthropology." To be fundamental, he argues, the human sciences cannot treat *anthropos* apart from the environment. In one of his earliest and most "existential" works, *L'Homme et la mort*, Morin was seeking to show that "there is no wall between nature and culture, but a meshing of continuities and discontinuities" (1951: 20). Human death, he maintained, is both fully biological and fully cultural. In a Hegelian turn of argument, he held that awareness of death brings our species to self-consciousness and launches its cultural vocation of constructing myths, religions, and systems of belief.

By the 1970s, when Morin had absorbed the systems approach, he extended this project by writing a virtual epistemological manifesto of French ecologism (1973: 211):

Fundamental anthropology must reject any definition that makes man either a supra-animal or strictly animal being; it must recognize man as a living being while distinguishing him from other life forms, it must transcend the ontological alternative nature/culture. Neither pan-biologism nor pan-culturalism, but a richer truth, which gives a larger role to human biology and culture, *because it is a role entailing reciprocal influence of one on the other.*

Arguments from *Le Paradigme perdu: la nature humaine* (1973) demonstrate how comprehensive Morin intended fundamental anthropology to be. He begins by showing how certain features of primate societies—relations of rank and subordination, the occasional use of tools, expressions of tenderness—were the founding conditions of human community (38–54). Using the language of the systems approach, he then traces the evolution of our "hypercomplexity," which integrates bio-physical structures (e.g., the human brain and central nervous system), individual behaviors, cultural adaptation, and ecosystemic conditioning. Each of these systemic levels is both affected by and affects the others. As a biological organ, the human brain is endowed with an aptitude for the integration of culture; it contains structures that are activated only by sociocultural education. Thus, certain bio-physical characteristics of the human brain are preconditions of culture. Over the course of human evolution, however, our ability to adapt to environmental change has depended on diminishing the influence of limited instinctual responses and on developing more complex "organizational competences." By developing cultural memory, by increasing technical knowledge, and by creating new behaviors and social rules, humanity has managed to adapt itself to a greater range of environmental conditions than any other species. Those very adaptations have even altered the physical characteristics of the brain. The brain needs time to assimilate culture, and (since culture is the key to mankind's adaptive successes) evolution favors genetic mutations that increase the length of time it takes for the brain to reach maturity (ibid.: 95–98).

Morin's chosen word for this complex process of reciprocal interaction of biology, culture, and environment is "hominisation." It is a process of "definitive incompleteness" because of our radically creative power to increase diversity and complexity (ibid.: 103).

The Lost Paradigm is the prolegomenon to a multi-volume work, *La méthode*,[3] in which Morin lays out the reflexive strategies for studying complexity in every domain. Crucially, his method emphasizes the existence of ontological borders between systems of different levels of complexity. The behaviors of entities on different sides of such borders require different explanatory principles. The laws of physics determine the behavior of brute matter. But when matter gets organized into a genetic program, it constitutes a life form that behaves teleologically and is able (within limits) to select matter and energy from its surroundings so as to maintain its form. The "program" that allows biological entities to do so succeeds by strictly regulating the functions of the organism's component parts. Organisms are, in a word, centered. Their internal genetic program gives them an orientation toward what is valuable in the world for them.

At a higher level of complexity, life forms interact ecologically. They exchange information, they program one another, they meet one another's needs as individuals or as species. This level of complexity has a different logic than the complexity of a life form. Ecosystems do not subordinate their constituents to a central program. "Eco-organization," explains Morin (1980: 238), "maintains itself and keeps itself in being, but lacks all self-reference and ecocentrism." Ecosystems are both "polycentric" and "acentric" (ibid.: 36). They are polycentric in the sense that they integrate and shape the activities of multiple species whose identities nonetheless remain distinct. They are acentric in the sense that they are in a continual process of reorganization. Morin emphasizes all that is contingent and unstable in the operation of systems. "The ambiguities, uncertainties, and 'noise' of the environment pose questions, problems, puzzles, charades to living beings who, in response, develop communicative networks that weave together in the ecosystem and thereby contribute to the enrichment of ecocommunication." (ibid.: 39) The higher the level of complexity, the less applicable is the idea that a living entity has a center. For Morin, ecosystems are not characterized by a stable form. There is no recurrent equilibrium point that might apportion significance to their component parts. It is their capacity for "eco-reorganization," their ability to "invent new reorganizations on the basis of irreversible transformations that occur in the biotope or biocenosis" (ibid.: 34–35), that challenges human observers to develop a correspondingly complex mode of ecological thought

Hominisation occurs on the far side of another ontological border. In the words of one of Morin's interpreters, "human societies are inscribed in natural history at the same time that they break with that history" (Rudolf 1998: 121). The emergence of self-consciousness gives the human subject an unprecedented degree of individuality and freedom. At the same time, it makes that subject profoundly dependent on a social milieu for its mental development. It is the cultural coordination of individual activities that has given humanity enough collective power to subjugate [*asservir*] every other species. Whether considered as individuals or as social beings, however, human beings never truly break free of their biophysical environment.

Morin teaches that ecology is not just a natural science. "Human societies have always been part of ecosystems," and "ecosystems, ever since the universal development of agriculture, animal husbandry, sylviculture, cities, are part of human societies, which are part of them. General ecology therefore must integrate the anthroposocial sphere into the ecosphere" (Morin 1980: 70). It is because we have not yet ecologized our thinking in this way that we face a "polycrisis." The predicament, as described in *Terre-Patrie* (Morin and Kern 1993: 108), "manifests itself through . . . increasing uncertainties, through ruptures in . . . negative feedback, by growth of positive feedback . . . , through the increase of dangers. . . . When we consider the state of the planet, we see: the increase of uncertainties in all areas . . . ; demographic growth, uncontrolled development of industrial growth and of techno-science; mortal dangers for all of humanity, including nuclear weapons and threats to the biosphere." Morin sees Earth's self-regulating systems in danger everywhere. His diagnosis identifies two factors in combination as particularly destabilizing. First, "techno-science is the kernel and the motor of planetary agony" (ibid.: 102). Society is organized according to the logic of an artificial machine, with a relatively simple "program," composed of highly specialized parts, carrying out certain strict functions (e.g., maximizing production). Lacking the adaptive complexity of a "living machine," the artificial machine amplifies destructive trends rather than inventing new behaviors to regulate threats. When confronted with hostile environmental conditions, it breaks down rather than regenerates itself. Second, the globalization of commerce and communication has spread the artificial machine everywhere, so that the machine's dysfunctions threaten to destroy the remaining sources of life-renewing diversity.

In response to this multifaceted crisis, Morin advocates a "politics of planetary responsibility" (Morin and Kern 1993: 166). Truly complex political thought would beware the myth of "the conquest of nature-object by man, the subject of the universe" (ibid.: 107). It would put brakes on the application of technical knowledge to nature and human cultures. Caution, not technological daring, should become our "global principle." Complex thought would—much as Passet advised—"subordinate the economy to a politics of man" (ibid.: 121). And nothing less than a planetary federation could deal adequately with our global polycrisis. Only internationally coordinated action could regulate growth and competition in ways that assured the survival of biological and cultural diversity (ibid.: 136). Although such a program would put limits on the sort of activities we now pursue in the name of human interest, however, it is not anti-humanistic. In fact, Morin insists that his program consists in "the development of human beings, of their mutual relations, of social being . . . —the continuation of hominisation" (ibid.: 166). Morin's multidisciplinary study of the dialectical interrelations of nature, culture, and disciplinary knowledge itself aspires to "the conscious control of knowledge by knowledge" (Bianchi 1990: 23).

Morin's "ecologized thought" undoubtedly resembles the philosophy of any number of thinkers versed in a systems approach. Several themes, however, set his work apart, both from English-speaking ecologism generally and from the ideas of some of his French colleagues.

In relation to English-speaking ecologism, Morin never pushes systems theory in the direction of a new, nature-centered environmental ethic. Indeed, his "organizationalist" interpretation of the systems approach constantly interferes with such a move. His insistence that human-caused environmental changes occur *within* ecosystems implies that human impacts on ecosystems are not inherently wrong. Actually, no serious environmental ethicist would disagree. But nonanthropocentric reasoning does depend on making a distinction between environmental changes caused by human beings and those imputable to other causes. Only changes caused by humans are morally suspect (Katz 1997: 43). As soon as environmental issues are framed in that way, it becomes nearly impossible to provide a persuasive, principled explanation of the range of morally permissible human activities. To acknowledge that human life depends on consuming nonhuman life, that reshaping the environment is integral to human cultures, and that humans have powers of judgment that distinguish them from all other species is to recognize why the project of devising moral rules ostensibly applying across all of "nature" regularly pulls up far short of its objective. An "ecologized thought" that does not begin by positing a moral distinction between humanity and nature need take no such theoretical detour.

Morin's notion that ecosystems are acentric also reveals the confounding blurriness of the supposed bearer of ecosystemic value. The American environmental ethicist Holmes Rolston III is as aware as Morin of the "looseness, decentralized order, and pluralism of biotic communities" (Rolston 1988: 183). And Rolston perceives how random events help drive ecosystemic evolution in ways that are identical with those that made Morin write of "acentrism." Rolston claims that holism leaves value "smeared out into the system" (1988: 188)—not radiating from a single point. But in order to make the case that specific duties arise in relation to ecosystems, Rolston has to pin down what it is that we must respect and preserve. In his holistic perspective (ibid.: 176), "the appropriate unit of moral concern" is less the individual organism than "the fundamental unit of development and survival." That is why "an ecologically informed society must love lions-in-jungles [and] organisms in ecosystems." Holism presupposes wholes. The whole is the fundamental unit that apportions value in supposedly determinate way. The whole operates as a moral center.

Yet Rolston also knows that "over evolutionary time ecosystems are quite historical" (1988: 178). Jungles change; some die away. In fact, such changes may be stages in the natural succession of ecosystems. Thus, perhaps our duties pertain not to this or that existing ecosystem but rather to the evolutionary process that diversifies life in general. That ethical position, however, greatly dilutes our duties, even to the point of insignificance. As Lovelock once observed, human activities are extremely unlikely to diminish the presence of earthly life to a point from which evolutionary processes cannot recover. It is simply life as *we* know it, especially our own, that our follies endanger (Lovelock 1995: 36–37). It is no wonder, then, that Rolston tends to correlate our duties to more narrowly identifiable "biotic communities." Valuing eco-centers comes at the cost of downplaying "acentered" processes that involve indeterminacy and creativity.

Just as Morin's organizationalism highlights the role of disorder in physical systems, so it acknowledges the ineradicable place of myth and irrationality in the human psyche (or, as Morin expresses it, in the history of *Homo sapiens demens*). For Morin, the human mind inevitably generates two strategic paths toward the apprehension of reality: the path of *logos* (empirical investigation, logical structure, and technical manipulation) and the path of *mythos* (magic, analogy, and symbolism). Myth's use of meaning-concentrating symbols and narratives constantly infiltrates the world of rational thought. Paradoxically, trying to purge reason of myth only makes us vulnerable to new and perhaps more dangerous myths. A member of the Communist Party from 1941 to 1951, Morin often reflects soberly on his own political gullibility. Some of the people who most prided themselves on critical thinking have been susceptible to messages of secular salvation.

An acknowledgement of the limits of rationality is important for purposes of countering anthropocentric uses of systems ecology. Unlike Rosnay, Morin criticizes cyberneticians who reductionistically try to express all systemic processes in terms of energy and information. Morin (1977: 238) maintains that cybernetic theorists have been singularly blind, and that they hide the "anthropo-social matrix of the cybernetic artifact: the power that . . . gives orders to the computer, that programs the program." Reconsider Rosnay, all of whose enthusiasm for decentralization seems to be at odds with his complaints that, through uncoordinated action, individuals have devastated their environment. The solution would seem to be to coordinate, program, and control. This solution implies the existence of programmers. And if programmers are to avoid reproducing the dysfunctional tendencies of the system's component parts, they must not be subordinate to those parts. They must have authority over them. Social programmers have power that is itself uncontrolled—hence the danger of authoritarianism.

Morin uncovers the deeper epistemological commitments that drive cyberneticians to think in these potentially authoritarian terms. They conceive communication as a matter of command, downplaying the place of disorder, conflict, and permanent reorganization in the constitution of living systems (Morin 1977: 251). A cybernetic artifact is an apparatus that transforms information into a program, thus into a set of organizational constraints. Through "servo-mechanisms," a cybernetic machine reregulates its actions as conditions in its environment fluctuate, thus allowing it to continue performing its function. In one sense, this "emancipates" the machine from environmental constraints. But in another sense, it presupposes that its constituent parts lose their autonomy. They become subject to the commanding parts of the apparatus. They are subjugated [*asservi*]—reduced to servitude. Applied to human societies, this logic suggests that the "mastery of nature" takes place through a historical process in which people have been reduced to subjection in a larger social order commanded by the state. Cybernetics, applied to society, can become the ideology of "technocentric, technomorphic, and technocratic" practices (ibid.: 252).

Thus, in regard to ecological questions too, Morin cannot conclude that we might aspire simply to eliminate the "magical elements in political discourse," as the Group of Ten's original question implied. That would be to fall into the errors of false rationality—a one-dimensional, compartmentalized notion of reason, and a notion that generates unintended consequences precisely because it is overly confident in its own powers. "True rationality," in contrast, "is open and engages in dialogue with a reality that resists it. . . . It is the fruit of an argued debate about ideas, not the property of a system of ideas. Reason that ignores real beings, subjectivity, affectivity, life, is irrational. It must allow a place for myth, for emotion, for love, for repentance, which must be considered rationally. True rationality recognizes the limits of logic, of determinism, and mechanism. . . . It negotiates with . . . the obscure, with what has been irrationalized and what is irrationalizable." (Morin and Kern 1993: 188) False rationality easily forecloses options because it glosses over uncertainties. Knowing little of the larger whole in which its proposed solutions will be implemented, false rationality foresees neither the environmental nor the human costs attached to them. For instance, detecting a need for higher levels of agricultural production, one may "rationally" prescribe the use of new fertilizers while ignoring how those chemicals will affect the region's ecology or how indigenous farmers may have cultural practices that work against the prescribed ways of dispersing them. Even a systems approach can reach the level of "complex thought" only if it can integrate culture and politics into its understanding of natural processes.

Ecologists can benefit from systemic understanding while checking its authoritarian penchant if they rethink the very concept of nature. The last and most crucial aspect of Morin's distinctive systems-theoretic project is its commitment to a form of epistemological reflection that tries to understand how concepts arise historically—even "ecologically" (Morin 1980: 93)—in a dialectic of physical influences, cultural practices, and subjective interpretations. Cybernetic models of "natural" systems are themselves products of such a dialectic. "We are machines," says Morin (1977: 285), "and at the same time, we are the ones who produce the concept of a machine." The history of science reveals that at times philosophers have ascribed human-like intentions to nature; at times nature has been conceived as a giant clock, as a competitive struggle for existence, and, cybernetically, as an immensely complex system of feedback loops, programs, and information flows. "We physicalize our notions then we socialize them, then we rephysicalize them and resocialize them, and so on." (Morin 1977: 285–286) Morin's "complex thought" not only attempts to grasp the multiple relations that make up what we call nature; it also tries to remain aware of the fact that the very concepts through which we conceive "nature" are caught up in this unending dialectic (Morin 1992: 72). From this perspective, even the best scientific model of "natural" systems cannot claim the sort of epistemic primacy that might justify giving social "programmers" unconstrained authority.

Morin is particularly helpful in getting us to see that intrinsic-value theory and systems theory, if they become too naturalistic, have an unexpected point of convergence: both take "nature" itself for granted. Both assume that it is relatively unproblematic to describe what nature is. The real challenge, for both types of theory, arises in how we *treat* that which we come to know as "nature." Intrinsic-value theorists urge us to "respect" it. Moderate anthropocentrists then urge us to "regulate" our activities so as not to destabilize delicate equilibria identified by systems ecology. Both are, in a sense, insufficiently self-conscious about what "it" is and about how human beings have come to perceive it as such.

It must be admitted, however, that there is a frustrating vagueness in Morin's prescriptions. What are the implications of "anthropolitics" for preserving biodiversity or controlling human population growth? On what grounds might we decide which technologies deserve to be treated with most "caution"? Do we really know how to subordinate the economy to a "politics of man" while preserving desirable forms of freedom and cultural innovation? Morin rarely descends close enough to the ground to answer to such questions. He is more a metatheorist of ecologism than an environmental ethicist or a strategist of green praxis.[4] What he provides are methodical reflections on transcending the nature/culture divide—reflections designed to make us adopt a political-cultural posture that would revitalize the spontaneous, self-regenerating capacities of the systems in which we are dialectically engaged. That posture takes advantage of the systems-theoretic insight that systems use forces of disorganization in their environment to evolve their own organization. In response to random events and disturbances, they create new means of detecting change in their surroundings and they generate subsystems to cope with such change. They increase their diversity and complexity. The growing complexity of human culture and technology is, in one sense, evidence of this very process. It has made the human species the most adaptively successful species on earth. At the same time, however, Morin warns that if our current adaptive strategies—universal exchange, high rates of energy consumption, monocultural food production—disrupt the self-organizing adaptive properties of the planet's biosphere, we may destroy the very stimulus of human creativity. Changing our priorities in favor of a "planetary anthropolitics" would not be a matter of respecting nature's intrinsic value or simply a matter of adopting a more enlightened understanding of human interests. It would be a matter of getting humanity to assume a posture that would allow a larger process of systemic complexification to continue to evolve in a direction that favors imagination, diversity, life, freedom, and renewal.

The Dialectic of Nature and Humanity

Luc Ferry opens his critique of ecologism by recounting a number of sixteenth-century trials in which animals stood accused of damaging human interests—and received legal representation that defended their right to life. Contemporary ecologism, with its occasional talk of animal rights and nature's intrinsic value, strikes Ferry (1992a: ix–xvi) as a throwback to premodern, prehumanistic modes of thought in which humans and nonhumans alike are submerged in a morally prescriptive natural order. To interpret the proponents of nonanthropocentric English-speaking ecologism that way was controversial, to say the least; to insinuate that pre-humanistic thinking was resurfacing in French ecologism too was simply bizarre. Had Ferry dwelt on the work of any thinker considered in this chapter, he might have seen that French ecologists, with their fascination for systems ecology, are more likely to be guilty of scientizing nature than of sacralizing it. Still, the prevalence of systems approaches makes the hardest test case for my thesis that French ecologism is humanistic through and through. That is why it is important to see that, aside from minor lapses, Passet, Rosnay, and Morin use the systems approach with considerable epistemological and political circumspection. Each understands that a new concern for systemic reproduction will take its place among a variety of longer-established goals of humane communities: goals of respecting political freedoms, self-governance, human rights, and cultural diversity. Most explicit, Morin (1990: 95) holds that "there is not a single ethics, but several, which can be required simultaneously and in contradiction to one another." His "planetary anthropolitics" ends by advancing not a definitive measure of ecological rectitude but a mix of "guiding ideas"—goals such as rerooting man in earth, preserving cultural diversity, and creating a community of nations. Even when speaking of the place of myth in the development of rationality, even when pleading for the protection of indigenous cultures against the spread of industrial society (Morin and Kern 1993: 91–93), Morin does not idealize "primal cultures" for their intuitive appreciation of nature, as some deep ecologists[5] do. His respect for scientific inquiry, and his commitment to probing the subjective origins of myths, steer him off that path.

What makes these French theorists humanists is their critical ethical pluralism, their recognition that no conceivable hierarchy of values gives any particular value (including preserving "nature") an a priori claim to take precedence over all the others. No major French ecologist forcefully presses either the claim that an ecological politics is one that would have to be guided by scientific understanding or the claim that "natural" things have intrinsic value such that altering them is presumptively morally suspect. French ecologists maintain humanistic traditions of tolerance and open-mindedness.

In its most philosophically ambitious manifestation, such humanistic open-mindedness appears as a developmental achievement in the dialectic of humanity and nature. That, at least, is the view of Edgar Morin, "the most Hegelian of all thinkers in the arena of systems theory" (Kofman: 1996: 57). Like Hegel, Morin regards "rationality" not as a timeless property of the individual mind but as a quality of human culture. Reason becomes more comprehensive as it develops historically out of dialectical confrontations between order and disorder and between pretensions to universality and eruptions of particularity. "Rationality," says Morin (1990: 261), "is the dialogue between the mind's will to logical organization and the empirical, external world." In Morin's hands, a "system" becomes a set of relational properties that transform themselves through crisis. In the process, they gain performative competence, autonomy, and eventually consciousness. History is neither a random succession of events nor a fated unfolding of a human capacity to understand and control nature. It has some systemic structure, in which random disturbances and "irrational" myth are integrated in the evolution of meaning. For Morin, "the earth is not the sum of the physical planet, plus the biosphere, plus humanity"; it is "a complex, physical-biological-anthropological totality, in which life emerges from the history of the earth, and man emerges from the history of terrestrial life" (Morin and Kern 1993: 69). "Humanity" itself, in Morin's eyes, is a historically emerging ethical sensibility, a growing awareness— and affirmation—of the unity of our species and its dependence on the earth.

Yet even a systems approach manifests only a moment in the development of rationality, not its definitive realization. Morin's history cannot be identical with Hegel's (Fages 1980: 226). Morin draws explicitly on the skeptical humanist tradition in French thought.[6] He reminds us that Pascal (as quoted in Morin and Kern 1993: 211) regarded man as both "judge of all things" and "imbecile worm of the earth." A skeptical humanist takes heed not only of the place of rupture and contradiction in human relations but also of the dangerous tendencies of hubristic human thought. Like Pascal, Morin is a critic not only of scientific reductionism but also of holistic arguments that the elements of a system have no significance apart from the whole. Morin recognizes that evolution is riddled with contingency and

ineradicable irrationality. History has no necessary end, no totalizing culmination in which all that exists will make sense and all that came before
will find its justification. Understanding entropy as the eventual fate of
every closed system, Morin can describe his ecologistic values only in terms
of "hope" for a reprieve from the gathering forces of disorder in our world.
Noncentered reason knows no necessity.

4

Politicizing Nature

Could there be a politics of nature—not a politics *about* nature; a politics *of* nature? Surely not, one is tempted to respond. Only people have politics. Only people organize themselves according to a vision of a good life. People argue, persuade, mobilize, and negotiate. Landscapes, animal species, and ecosystems do not reflect on their own interests, do not develop strategies of control, and do not seek to legitimize their rule by making those affected by their power see their status as inevitable or desirable. In this area, the nature/culture boundary is insuperable. We may agree with William Ophuls that "the duty of the next generation of philosophers [is] to create an 'ecological-contract' theory promoting harmony not just among humans, but also between humanity and nature" (Ophuls and Boyan 1992: 215). An ecological contract, however, can only be a matter of making nonhuman things the *objects* of a more environmentally rational human politics.

Now consider matters from a different angle. The language we use to describe politics—human politics, we suppose—crosses the nature/culture boundary every day. It is not only words such as 'power' and 'violence' that apply in both "realms" (*royaumes*—territories ruled by kings). Humans and nonhumans alike can have a "constitution," a form. They are said to obey "laws." Cyberneticians talk of nonhuman processes as "governing" themselves. There are plant and animal "communities." What if we found not only that such conceptual crossovers are common but also that the terms used to describe "human" politics were, *from their inception*, often developed out of observations of nonhuman phenomena? What if "law," "contract," and "power" had no pure origins in the experience of human communal life? Might it not then make sense to say that there is something political in what has traditionally been called "nature"?

Ordinarily, politics pertains to the processes of persuasion, competition, and issue framing through which groups with differing values achieve sufficient reconciliation to constitute a community. In a community, matters of common concern appear in a public realm. Rights get defined and protected, and tangible goods get distributed to various constituencies in the community as a function of some shared understanding of a good life. When I say that Michel Serres politicizes nature, I allude to his account of interchanges that constantly mix such "political" activities and ways of understanding and manipulating the surrounding world. What "nature" is and how we see fit to relate to it (e.g., as conquerors, inhabitants, co-equal contractors, or cautious experimenters) are written into the terms in which we describe both our communal life and our nonhuman surroundings. At the same time—symmetrically, Bruno Latour will say—communities take on attributes learned from their interaction with the nonhuman world. These theorists thus go beyond the French thinkers discussed so far in the degree of their epistemological radicalism. They refuse even a dialectical understanding of humanity's interaction with the material world. For them, conceptions of right, law, community, and representation run as much through our conceptions of "nature" as they do through human institutions. Authority and power are inscribed across the intermingling domains of nature and culture.

Symmetrical study is an especially French way of dissolving the center of political ecology. Sometimes (but too rarely) English-speaking ecologists start from an acknowledgement that "nature" will always be a culturally relative, politically contested, value-laden term (Barry 1999a; Macnaghten and Urry 1998). It is important to see, however, that the French theorists considered in this chapter would not even concede that green theory is ultimately "human based" (Barry 1999a: 215). In a symmetrical study, "human" characteristics are so interwoven with "natural" ones that such an assertion is no longer deemed meaningful.

English-speaking aficionados of recent French philosophy also should be warned of another peculiarity of nature's politicizers. It would be easy to assume that contesting unwarranted "naturalizations" is a form of postmodern criticism. "Politicizing nature" might be thought to refer to power struggles between groups seeking to encode their preferred modes of social control and discipline into discourses concerning nature. But postmodernism has been less influential among French ecologists than one might expect. The rarity of centered ecologisms in France deprives postmoderns of their most obvious targets. Indeed, many French ecologists explicitly criticize postmoderns for being insufficiently attentive to the actual processes of scientific investigation that so profoundly shape our views of nature. It is the reflexive realism of noncentered ecologisms that postmodern analyses of codes and discourses tends to miss.

Serres on Science, Structuralism, and Violence

Michel Serres is an historian of science trained in mathematics. Elected to the Académie Française in 1990, he is one of the most prestigious intellectuals in France today. He is also the French intellectual most likely to be identified by commentators with ecologism. This is mainly due to his 1990 book *Le contrat naturel*. Abundantly reviewed in the press and earnestly cited by leading figures of Les Verts, that book seemed finally to align ecologism with French traditions of the *intellectuel engagé*. Like Voltaire or Sartre, Serres is a writer of international renown who decided to intervene publicly, controversially, and with philosophical panache in one of the most urgent questions of his era.

At one level, *The Natural Contract* is about a crucial deficiency in the social-contract tradition. According to that tradition, government is the result of an agreement among potential citizens to distribute the benefits and burdens of cooperation in a way that they deem fair. The social contract replaces private judgment and vengeance, thereby reducing violence. But, says Serres, the environmental threats that humanity now faces reveal the dangerous partiality of the *social* contract: it treats the *natural* world as a field of passive objects available for appropriation and destruction, thereby threatening the equilibria on which life itself rests. In a passage suggesting his debt to the systems approach, Serres (1990: 66) formulates the challenge this way: "There are . . . natural equilibria, mechanics, thermodynamics, the physiology of organisms, ecology or systems theory. Similarly, cultures have invented . . . several human or social equilibria, that have been chosen, organized, protected by religions, law or political arrangements. We have failed to conceive, construct and to implement a new global equilibrium between these two wholes." Serres argues that we must learn to negotiate a new contract with a nature whose links of inter-dependence and equilibria generate rights that we must respect. Reciprocity with nature must replace mastery over it as the regulative norm of human activity (ibid.: 67).

The natural contract includes "nature" in a planetary sense. The social contract was always implicitly local. In spite of the abstractly universal character of their imaginary contract, theorists envisioned it as legitimiz-ing something like a national community—a group of people probably sharing a language, a limited territory, and certain cultural traditions. The natural contract, however, pertains to the planet as a whole and to future generations. What is unprecedented in our ecological predicament is that powerful technologies of energy production, communication, and com-merce, as well as the instruments of war, are affecting environmental qual-ity world-wide. We have created a "world-object." The natural contract must be correspondingly global and incorporate the interests of future gen-erations (Serres 1990: 67–73).[1]

What should we do, however, when scientists themselves are not certain about the global effects of human activity on the world? For example, what if it cannot be proved definitively that emissions of greenhouse gasses are causing global warming? Shouldn't we await more certain evidence before acting? Serres responds by borrowing a page from a skeptical humanist. Blaise Pascal reasoned that it made sense for individuals to wager that God and the afterlife did exist, even if they had doubts about them. Eternal life and happiness are the highest of stakes. If one lives a godly life and God exists, there is everything to gain; if God does not exist, "you lose noth-ing," claimed Pascal (1971: 439). For Serres, identical reasoning pertains whenever there is potential for global environmental disaster. Suppose a danger is alleged to exist. If we bet against its materializing, and we win, life goes on as before. But if we lose, we lose everything. Now suppose that we respond to the danger by taking action to avoid it (e.g., reducing emis-sions of greenhouse gases). If we are wrong, "we lose nothing." But if we are right, "we win everything and remain actors in history" (Serres 1990: 19). Where the highest goods are in jeopardy, caution is the order of the day.

Ferry warns that Serres has begun a process of smuggling anti-humanist, undemocratic, and disturbingly un-French notions of nature's intrinsic value into the homeland of Cartesianism. The notion of a "natural contract" is philosophically unrigorous, he objects, because "in a humanistic juridical framework, nature can have no status other than that of an object, not a subject" (Ferry 1992a: 240). Nature is not an intending subject; thus, it cannot have rights and it cannot negotiate a contract in any literal sense. In Ferry's interpretation (ibid.: 150–151), Serres's point is that "man must go from being a 'parasite,' which . . . conducts relations with nature in an inegalitarian way, to becoming a 'symbiote.'" Ferry grants that Serres's notions of contracts and symbiosis might be read as rhetorically captivating metaphors. If they are metaphors, however, they are not really justifications for recognizing nature's intrinsic value (ibid.: 244–245).

It is true that Serres talks of treating nonhuman things as "subjects of right." However, by reading *The Natural Contract* through the lens of English-speaking ecologism, Ferry misses all of Serres's arguments that place his assertion of rights far outside the rhetorical boundaries ordinarily observed by environmental ethicists. Serres does not derive nature's "rights" directly from the subject-like properties of natural organisms (e.g., functional wholeness) or ecosystems (e.g., goal directedness), as many intrinsic-value theorists do. He feels free to adopt the language of contract, law, and rights, not because he believes that nonhuman things *have* rights, but because our languages for describing nature *and* politics have, for millennia, drawn on the same pool of concepts and relations. Political theorists have made "nature" the standard of right, and naturalists have (with varying degrees of awareness) written political relations into their views of nature. It is this symmetry between political culture and science that permits Serres to travel across supposedly impassable divides between the human-juridical order and the objective-natural one. To understand *The Natural Contract* as a work of French ecologism, then, one must put it in the context of a philosophical project that Serres has been pursuing since the late 1960s.

Serres has written more than twenty books. They cover an encyclopedic range of topics: interpretations of Émile Zola and Jules Verne, meditations on the art of Carpaccio and on the physics of Lucretius, discussions of many major figures in the history of philosophy, explorations of communications theory, thermodynamics, and topology.[2] Serres does not indulge in aimless eclecticism. He practices a form of philosophy that Bruno Latour calls

"Enlightenment without the critique." "Critical" interpreters dominate the texts they analyze; they develop powerful commentaries to explain the authors' strategies of emplotment, the psychology of their characters, the emotive effects of their rhetoric, their political ideologies. Critique does not simply place texts in context; it *replaces* the text. Critical interpreters presume their frameworks to encompass truths that govern what is really taking place in the works they study. But Serres refuses to set philosophy up as the metalanguage which decides the truth of one domain of knowledge or the other. He juxtaposes literary and scientific texts without setting himself up as their master, without judging that one genre of discourse outranks another (Latour 1987a: 85–87). His studies of crossovers between scientific and humanistic expression leave epistemology "pluralized and relativized" (Serres 1972: 32) His theory of knowledge precludes pretensions to intellectual hegemony in any area of knowledge.

In a study of modern French philosophy, Vincent Descombes (1980: 86) categorizes Serres as a structuralist, putting him in the company of the anthropologist Claude Lévi-Strauss and the semiologist Roland Barthes. Structuralist analyses show how, in texts and in cultural practices, meaning arises out of the patterned arrangement of elementary parts, whose nature is left unspecified. Meaning is not inherent in the parts themselves; it is a result of their structured relationships. Serres approaches texts by abstracting away their specific content and detecting how their concepts and themes form relational networks. Such abstraction permits him to find structural analogies between works whose explicit meanings may seem wholly dissimilar. He demonstrates that writers in the most unrelated genres and writers separated by millennia nonetheless express similar and valid insights.

Long before Serres made any specifically ecological pronouncements, his structuralism recast the problem of connecting nature and culture. In one famous essay, he examined the ways that Zola's novelistic treatments of genealogy and family relations contain the basis of nineteenth-century genetics and, ultimately, thermodynamics (Serres 1975). Recall that systems theorists regard thermodynamics—the branch of physics that studies the laws of conversion of energy—as fundamental to understanding how complex systems function. For Serres, thermodynamics cannot be seen simply as a theory *of* nature. It uses a model of relationships found throughout the cultural formation of "the modern age." In the modern age, the steam-driven engine has replaced the "fixed point" as the organizing model of understanding. Descombes (1980: 90) calls Serres "the virtuoso of the isomorph" for his skill in finding formal equivalences across such disparate domains of knowledge.

To understand the isomorphisms of which *The Natural Contract* is constructed, we must turn our attention to the first mention of that concept in an essay written more than 10 years earlier. In *The Birth of Physics in the Text of Lucretius* (1977), Serres maintains in all seriousness that a didactic poem written in the first century B.C. contains a view of nature that is structurally homologous to that of a modern physics inspired by thermodynamics and cybernetics.

In Lucretius's resolutely nonteleological philosophy, the universe consists of randomly falling atoms. It has no intrinsic order. All is flux, dispersion. How then do recognizable bodies take form? Structures form through the phenomenon of the *clinamen*. A tiny deviation in the monotonous flow of atoms creates a turbulence in which they connect and spiral together in a temporary order. Lucretius describes the emergence of something by deviation from a state of equilibrium. Stability is now homeorrhesis—not a solid structure but the constancy of a flowing, localized circuit of matter perceivable against a background of formless flux (Crahay 1988: 78). "The Lucretian world," says Serres (1982: 116), "is globally entropic, but negatively entropic in certain swirling pockets." The clinamen begins a process of self-conservation. "Repair, feeding, feedback" are the productive phenomena that constitute natural forms capable of resisting the eternal flux (Serres 1977: 76). Lucretian physics posits an empty space and a flow of atoms as the basic reality of the universe; they are the "foedera fati," the laws of fate. The forms that take shape from the clinamen, born in radical indeterminacy, cannot be attributed to fate. They are "foedera naturae"— natural contracts. They have enough stability for their laws to be described, but they are too fragile, local, and variable to be called fundamental laws of the cosmos.

Still "science remains science and laws remain laws but what changes is the global contract, the general scheme of things that scientists agree to call 'physics'" (Serres 1982: 102). Contemporary scientists have begun to formulate a physics that is structurally akin to Lucretius's physics of flux and turbulence. The very title of a systems-theoretic book that is influential in many ecological circles, *Order Out of Chaos*, could have easily been adopted by Lucretius. The "nature" of Isabelle Stengers[3] and Ilya Prigogine is dynamic, processual. They argue that living things (and some nonliving ones) form "dissipative structures" in an entropic universe. Such structures strive to maintain their form by regulating the flow of energy though them and metabolizing other systems' structures. When random environmental fluctuations overwhelm those structures, more complex systems emerge to create new zones of stability. Indeterminacy, flux, emergent structure, localized stability, entropy: these themes of modern physics are the themes of Lucretian atomism too.[4]

Serres is not saying that Lucretius got nature right 2000 years ago and that modern science from Descartes to Laplace got it wrong. Like other French ecologists, Serres is engaged in *reconceiving* the nature-culture nexus, not simply correcting our view of nature. This early discussion of the natural contract shows how this is so. Serres turns to Lucretian physics because, unlike twentieth-century science, Lucretius offers a conception of nature that is not complicit in the violence of warfare. Though many philosophers since the Enlightenment have seen in modern science the path to human reconciliation, Serres emphasizes throughout his works the greatly enlarged capacity for violence that they create. Referring to the development of the atomic bomb and its relationship to "scientific optimism," Serres writes: "Hiroshima remains the sole object of my philosophy" (1992a: 29). The bomb over Hiroshima obliterated any illusions that science could only be a force for life and ethical progress. Serres ends his essay on Lucretius with a dark allusion to this central theme (1982: 124): "Violence is not only in the use of science but still hides in the unknown of its concepts. . . . The world after Hiroshima can still die from the atoms."

"The unknown of its concepts" refers to Serres's argument that science proceeds from certain founding images and metaphors that condition the ways—violently or lovingly—it handles the phenomena it studies. One physics takes shape under the sign of Mars, the god of war. It is a physics of rank-and-file formations, parallel lines, chains, sequences, rivalry, power, and competition; it is a view that imagines "cutting up the bodies into atomized pieces, letting them fall." This is "the foedus fati, what physics understands as a law; things are that way" (Serres 1982: 100). And then, with a political extension of his argument, Serres (ibid.) notes that "it is also the legal statute in the sense of dominant legislation." Serres investigates how a strategic and political terminology—the martial metaphors—shaped understandings of the "natural" world from the very start of scientific understanding and how those understandings were transmitted historically in the works of Francis Bacon and René Descartes.

But there is an alternative. Lucretian physics forms in the presence of Venus, the goddess of love. This is a physics of "vortices, of sweetness, and of smiling voluptuousness," one in which there are flows, turbulence, and a degree of freedom. In the Lucretian view, order is achieved not by domination but through a gathering together, a conjugation of things that are in constant, vortical movement—a foedera naturae, a natural contract. Here "nature is formed by linkings, . . . relations, crisscrossing in a network" (Serres 1982: 114). Serres's point is that science is always "conditioned" by founding concepts drawn not from science itself but from other domains: politics, military strategy, mythology. These founding concepts do not determine the content of scientific knowledge (e.g., the law of gravity or of entropy in closed systems). However, they do affect what Serres calls the "topography" of the sciences: the "global form [of a science's parts] and its relief at local points." And these in turn influence how the science is used. "The physics of Lucretius . . . is in fact the same as that of Archimedes, but the postulation of Venus and the exclusion of Mars transform it. Hydrostatics in the first is related to the constitution of living beings; in the latter it is related to the theory of ship-building." (ibid.: 107) An understanding of the world informed by notions of fluidity and change rather than order and repetition, by contractual federation rather than strategies of domination, is, Serres believes, more consistent with the flourishing of life in general.

In this essay, then, Serres does not see the "natural contract" as an ethical act in which people come to an agreement to respect all the values they attach to nature. Nor does he contend that the natural contract comes from a perception of nature's pre-existing values, which then receive juridical recognition. The natural contract is a concept that arises from an attempt to conceptualize phenomena in a particular way in which "knowing" means understanding the bonds and interactions that create form-sustaining structures in an essentially disorganized universe. Though our socially inherited ways of thinking are shot through with notions that predispose us to adopt attitudes of violence and mastery in relation to our world, contractual notions are available with which we might establish a more life-sensitive "topography" of existence.

The Meaning of the Natural Contract

In *The Natural Contract* Serres continues the line of argument laid out above. Essentially, he exploits structural equivalences between contractual understandings of human communities and scientific understandings of nature in order to allow us to conceive an ecologically responsible politics without getting caught in the nature/culture dualism. The equivalences cluster around two related concepts: contract and law.

Let us consider contract first. The kind of assertion that disturbs Ferry occurs when Serres (1990: 69) announces that "the earth speaks to us in terms of forces, connections, and interactions, and that suffices to make a contract." Ferry (1992: 151–152) retorts that this is inconceivable, insofar as contracts presuppose autonomous subjects capable of making moral commitments, and that Serres can speak of a natural contract only by misleadingly ascribing human properties to the earth. Serres would point out, however, that contracts are not so firmly associated with subjectivity. The original image of "contract" is more physical than metaphysical. The word comes from *com-trahere*, to draw together. It conveys an image of tightening ropes, as in adjusting the rigging on a sailing vessel. If we speak of contracts today as moral and legal bonds created by bargaining among free individuals, something of the physical founding image still remains: a complex set of linkages, combining constraints and freedoms, in which each element receives information through every adjustment in the system as a whole (Serres 1990: 162). The language of contract does not presuppose a metaphysical divide with autonomy on the one side and physical necessity on the other. Contract is therefore more than a foundational concept in relation to the origin of human communities. Serres reveals the forgotten role of contract in imagining relationships between human and nonhuman beings in terms of equilibrium, partnership, and reciprocal influence. By mapping out formal equivalences between humanity and nature, we can begin to see the possibility of a "negotiation" with nature in which certain humane practices that have brought peace to human communities are applicable to nonhuman ones.

Law, even more obviously than contract, traverses our understandings of nature and culture. Serres (1993: 106)[5] once explained to an interviewer how the argument of *The Natural Contract* grew out of his curiosity about a thematic continuity in two seemingly distinct discourses:

> It was a question of knowing . . . why the term 'law' is used to designate certain scientific facts (laws of optics, laws of falling bodies, etc.) and this or that juridical disposition. It is this desire to take in and comprehend with a single glance the foundation of science and of law [*droit*] that gave birth to such problematics as those of natural right in Aristotle, or modern natural right. . . . My work consisted in . . . pursuing the question of whether it is possible to hold together juridical obligation and scientific rigor. In other words, aren't science and law part of a single foundation?

It is commonly thought that legal authority derives from the consent of a community of free citizens. Scientific authority rests on rigorous experimentation, observation, measurement, logic, and replication of results. Citizens and jurists have no more standing to judge scientific claims than scientists have, qua scientists, to determine the validity of law. Serres (2000: 21) observes that science is far more contractual than this division suggests. First, science and social-contract theory have a common strategy for overcoming clashing opinions: each gets disputants to distance themselves from their personal inclinations and to agree to submit their opinions to a generalizing viewpoint. According to Serres (1990: 79), "the scientific contract manages, cleverly, to make us take the point of view of the object . . . , just as other contracts . . . , through the ties of obligation, make us take the point of view of the partners of the agreement." When engaging in negotiations for a social contract, the parties cannot expect the adoption of rules proposed solely for the benefit of individuals in their particular situation; they must frame proposals likely to be acceptable to the group as a whole. In effect, they adopt a vantage point that allows them to determine a common good. Likewise, to engage in science is to renounce the right to enshrine one's predispositions, prejudices, religious beliefs, or interests immediately as truth. Claims about the nature of the world must be subjected to methodical scrutiny before winning approval. Second, even scientific validation requires a contractually linked community. To be recognized as scientific, knowledge must emerge from accredited practitioners who use conventional terms and accepted procedures. Those practitioners' results are then tried before juries of scientific peers. In this sense, "science proceeds by contracts" (ibid.: 42).

But it is not only contracts internal to the scientific community that validate its knowledge. Serres insists that confrontations between scientists and the larger civic community are integral to the history of science. That is, trials such as those of Galileo and Lavoisier should be seen as something other than moments when the forces of obscurantism blocked the march of reason. Serres contends instead that scientists' appearances before such tribunals are the moments when the internal history of the sciences (the scientists' own judgments of truth) and their external history (the community's judgments of right) get synthesized: "Truth, which seems to us like it must be founded on something other than an arbitrary convention, is on the contrary so founded. . . . What is science, knowledge, even thought? The totality of confrontations with all the other foundations of truth with this fundamental act of arbitration. Every certainty must present itself, to be registered and confirmed, to be canonized, before a tribunal." (Serres 1990: 121–122) Like Lucretius's atoms falling in a void and federating in natural contracts, knowledge, according to Serres, takes shape against a background of chaotic noise and violence. It forms in the struggles of different groups to vindicate their own understandings of truth and to obtain social recognition of their contractual protocols for getting access to it.

In this struggle, science has proved extraordinarily successful. Since Galileo's trial, external tribunals have dared less and less to call science to account. Against the Church, Galileo claimed the right of scientists to speak for things themselves, to describe and manipulate the world as they saw fit. Serres treats Galileo like the originator of property in Rousseau's *Discourse on Inequality*: one who marks off a territory and finds others "simple enough to believe" that their traditional rights would survive unchanged (1990: 133). Only later—too late—do they discover that their rights do not survive. Traditionally, civic and religious authorities have claimed the right to judge all matters pertinent to the community's well-being, often in the name of "natural law" or God's ordinances. In the modern world their writ runs no longer. "Nature becomes a global space, empty of men, left alone by society, a place where the scientist judges and legislates, that he masters, a place where positive laws leave technicians and industrialists almost in peace. . . ." (ibid.: 134) At least they are left in peace until the technologies they invent impose global environmental dangers that civic authorities cannot ignore. For Serres, the community of citizens needs to reassert the authority that it unwittingly abdicated.[6]

Serres (2000: 14; 1992a: 251) describes the challenge currently facing mankind as "the mastery of mastery." Especially in the latter half of the twentieth century, we learned to master the things that had kept humanity in subjection since time immemorial. The sciences have given us truly effective ways of dealing with problems of workplace organization, disease, sexual failure, short life expectancies, and famine. "We have become," Serres asserts (1992a: 248–249), "masters of the great figures of our ancient dependency: the earth, matter, life, time and history, humanity, good and evil." But learning the techniques of mastery has come at a great cost. Now, it turns out, everything is coming to depend on human choice. And it is this degree of mastery that eludes our mastery. As masters of the earth, we seem uncontrollably to destroy it. We know that whatever can be done technologically eventually will be done, in spite of many people's moral reservations. Serres has no doubt that the future is one in which "we are going to decide everything." The question that remains is not whether that will happen; it is whether we will approach decisions in the spirit of mastery or in the spirit of contract.

Unfortunately, Serres barely pursues the institutional implications of his ecological thought. How should the authority of citizens manifest itself politically: in legal limitations on genetic research in universities and industry? in communal regulations favoring "soft-energy paths" rather than the use of fossil fuels? in laws proscribing the production of chemicals implicated in climate change? This much we know: The natural contract would grant rights to some nonhuman things and would increase communal overview of scientific research and the diffusion of technology. But what things should get rights: individual nonhumans? species? ecosystems? Serres never details how the "natural contract" might be institutionalized; thus, he never faces up to potentially powerful objections.

Serres's defense of "the city's" right to regulate science opens profound questions about intellectual freedom. Wouldn't thoroughgoing politicization mean that virtually every form of scientific research, even basic research, could proceed only if given prior authorization? At what stage should the research that produced DDT have been made a matter of public debate: when the pesticide went into industrial production? just before field trials? Such late intervention hardly answers to Serres's concern that "anything that can be done will be done," so let us go further back in the chain of research. Perhaps the scientists who first designed laboratory tests for DDT should have been made answerable to some political authority. How could politicization have stopped there? Wouldn't it have been argued that whole lines of basic research in organic chemistry (and genetics, and meteorology, and nuclear fusion, and . . .) are too dangerous to be left to scientists? Had those arguments won, it is hard to see how scientific research could have proceeded at all. It is doubtful that that is Serres's intention.

Also unsatisfying is Serres's vagueness in regard to how to proceed in the face of choices that we now face every day. His version of Pascal's wager, for example, fails to foreshadow the higher ecological wisdom that he advocates. Apparently, Serres contends in regard to the danger of global warming that "if we choose responsibility" (i.e., if we see human activity as the source of climate change and hold ourselves accountable for curtailing emissions of greenhouse gases) the worst that can happen is we lose. But then "we lose nothing." For the sake of argument, however, suppose that ecologists are wrong and that human activities are having no significant effect on the earth's temperature. Then many people would say that undertaking vast reductions in emissions of greenhouse gases *does* entail losing something—perhaps a great deal. Cutting energy production could mean traveling less, producing fewer goods, turning off our air conditioners. . . . Huge expenditures would be necessary to modify agricultural practices all over the world. What makes decisions in this area so politically delicate is people's awareness of precisely such competing considerations.

Serres (1993: 106) brushes off practical objections by asserting that "the role of the philosopher is obviously not to conceive concrete applications." In his view, philosophers "advance by producing concepts." Fortunately, anyone searching for "applications" of Serres's ideas has recourse to an able interpreter and promoter of Serres's ideas who has a more practical bent: Bruno Latour.[7]

Latour's "Symmetrical Anthropology" of Science

Latour (1987a: 93) proposes that we regard the enigmatic philosopher Serres as a species of anthropologist:

Instead of believing in divides, divisions, and classifications, Serres studies how *any* divide is drawn, including the one between past and present, between culture and science, . . . between order and disorder. . . . Instead of dealing with a set, he always takes as the only object worth the effort the extraction of the set *from its complement.* . . . What would you call someone who chooses the extraction of the set from its complement? I call him provisionally an *anthropologist of science.* We are in the habit of thinking that anthropology's goal is to make sense of whatever non-scientific, pre-scientific, or anti-scientific beliefs and cultures there are left. . . . Studying how [Trobrianders or lower class Britons or Carnot or Prigogine] divide and order, studying what it is to *pertain* to something, this is the purview of an anthropology of science.

Anthropology signifies two things for Latour. First, anthropologists study "foreign collectives" comprehensively and holistically (Latour 1991: 24–25). They tie together patterns of material exchange, religious belief, marriage, political power, ancestors, cosmology, and property in a single narrative. They perceive common themes and structures across multiple domains of social practice. Second, of all the human sciences, anthropology comes closest to being consistently symmetric in its approach to its subject matter (Latour 1999b: 277). Latour credits anthropologists with devising methods for studying "foreign" cultures without supposing that their own culture is superior or more rational. Asymmetrical inquiry partitions "truth and falsity, efficiency and irrationality, profitability and waste . . . in two incompatible realms" (Latour 1994a: 791). It supposes that knowledge advances through rational procedures such as data-gathering, quantification, controlled testing, comparison, and logical analysis. Practitioners of symmetrical inquiry reject the epistemological mapping that places unreason and error on one side, science and truth on the other. Serres, explains Latour (1990: 161), is "a sort of symmetric Lévi-Strauss who would add to the diversity of primitive myths all of the scientific ones."

This approach to science is not Serres's alone. Latour adheres to the same methodological agenda. Latour is a philosopher and historian of science who elaborates on the program of symmetric anthropology by giving it a finer historical texture and pushing it toward a political ecology preoccupied with the problem of coping with unconstrained artifice (Mongin 1994:

352–356). His account of the processes through which "humans and non-humans swap properties" exemplifies more concretely than any theory since Moscovici's *Human History of Nature* how French ecologists try to move in a noncentered rhetorical space.

Latour's international reputation stems from his work in an area sometimes called "the sociology of scientific knowledge" and sometimes "science and technology studies." Latour now prefers to call it "the symmetrical anthropology of science." The different labels are an index of a progressive radicalization of his thought. His first major work, *Laboratory Life*, is best classified as an essay in the sociology of scientific knowledge.[8] Latour enters a laboratory to study the scientists who discovered a hormone regulating the human endocrine system. The hormone's presence—its very existence—is measured by a bioassay whose reliability scientists establish only by their persuasive abilities. What ends up counting as a "fact" about nature emerges only as a result of scientists' building networks of allies. An observer of such a process, says Latour, would "portray laboratory activity as the organization of persuasion through literary transcription" (Latour and Woolgar 1979: 88). But if scientists are not actually doing what they claim to be doing (discovering facts about the world), what motivates their work? Latour argues that they are engaged in a cycle of building up their credibility and reinvesting it in the "continual redeployment of accumulated resources" (ibid. 1979: 198). This motivational explanation is a variation on the ways that sociologists of science trace the influence of personal interests and economic forces on scientific inquiry. Latour's early work makes scientific "fact" appear as something constructed through social negotiation.

In subsequent books, Latour diminishes the role of human interests in the constitution of knowledge. Correspondingly, he accentuates the role of "actants." Actants may be people, but they may also be "natural" entities (e.g., bacteria), measuring devices, or machines. Latour maintains that the networks of humans and nonhumans out of which scientific facts emerge are so tangled that the distinction between subjective agents and natural objects breaks down. Louis Pasteur's extraordinary success in demonstrating the bacterial origin of anthrax, for example, was not, in Latour's view, the result of a genius' presenting incontrovertible evidence to scientific peers. His demonstrations required the mobilization of farmers, army doctors, hygienists, and statisticians, each of whom performed an essential part in creating the conditions under which Pasteur's claims solidified as fact. And these were not the only actants. Pasteur also enlisted *the microbes themselves* as allies, by "offer[ing] these ill-defined agents an environment entirely adapted to their wishes" (Latour 1984: 91.) In his flasks and under his microscope, Pasteur got bacteria to behave in ways that made them believable as the cause of anthrax. Latour is perfectly serious in presenting nonhumans as active participants in the networks that support our understandings of nature and society.

Bacteria as "agents" in their own society? This claim sounds peculiar, even perverse. But what makes it sound perverse, Latour would maintain, is our ill-founded, "asymmetrical" assumption that agency pertains only to human beings. If we abandon the assumption that what we now believe to be true was destined to become so, and if we investigate moments of controversy before the "facts" are established, then we find that our knowledge is the outcome of a vast array of processes, many of them contingent and "irrational." A crucial scientist may be an effective coalition builder or a social misfit. There may be powerful social groups prepared to disseminate a technology, or there may not. Whether these agents *intended* to solidify certain facts is not relevant to the status of facts as facts. And since "irrational" and unintended factors of these kinds influence "factual" outcomes, nonhumans too can count as equal participants in these processes. Bacteria act on the same stage as farmers, army doctors, and other scientific constituencies: all of them must perform in certain ways in order to translate Pasteur's ideas from idiosyncratic hypotheses to widely accepted truths.

Latour's works of the 1990s, especially his contributions to political ecology, take the principle of symmetry one step further. No longer is it a matter of refusing to prejudge the processes constituting knowledge as rational or irrational. Now it is humans and nonhumans generally that Latour proposes to treat symmetrically. Taking his cues from Serres, Latour undertakes a "pragmatogony," an account of the reciprocal constitution of human subjects and the objects of their world.

A pragmatogony differs importantly from social constructivism. Constructivism makes truths about the physical and biological world a function of social structures. Yet "most of the features of social order—scale, asymmetry, durability, power, division of labor, role distribution, and hierarchy—are impossible to define without bringing in socialized nonhumans" (Latour 1994a: 793). To be truly symmetric, we cannot accord causal primacy to society in the constitution of things any more than we can make "objective" nature the cause of social organization. Both society and nature are entangled "collective facts." In a pragmatogony, "Nature and Society are now accounted for as the historical consequences of the movement of collective things" and "all the interesting realities are no longer captured by the two extremes, but are to be found in the substitution, cross-over, translations, through which actants shift their competences" (Latour 1990: 170–171).

Latour suggests that political ecology itself is a product of "competences" shifted between humans and nonhumans. He argues that nature gets politicized when juridical relations elaborated in society are transferred to nonhuman things *and* when this transference is deemed plausible because those very relations were developed earlier in relation to things. So far, the argument is familiar from Serres's *The Natural Contract*—and it is not likely to sway critics who charge that the idea of transferring properties from humanity to nature creates metaphysical monstrosities.[9] Thus, it is significant that Latour takes Serres's philosophical insight and makes it concrete, more empirical, and less dependent on literary excursions and leaps of poetic intuition. Latour's innovation is to chart a strictly ordered process through which humans and nonhumans exchange properties and thereby form "collectives." His methodological principle is to search for an alternating pattern in which (in one stage) objects get enrolled in human relations by being endowed with social relevance and (in another stage) social relations get stabilized by acquiring properties from objects (Latour 1994a: 806).

Examples from two stages (Latour discerns eleven in all) clarify his procedure. Émile Durkheim and other nineteenth-century social scientists became aware that "society" (a corporate body transcending individuals) was molding them, giving them values, and slotting them into roles. Awareness of this highly structured human order, Latour contends, was itself the result of the growth of technical competence (that is, the ability to manipulate tools and matter purposefully). From techniques, "we learned what it is to last longer, to be spread over space and time, to occupy a role, to be dispatched to a function." The competences that organize people into "societies" thus stem from experience with things. But technique is not the ultimate origin of social structure; it too has a history. Technique arises through a type of human interaction not yet sufficiently formalized to count as "society" yet organized enough for people to learn to remove objects (stones, wood, animal skins) from their original settings, to recombine them (e.g., into a hammer), to use them in new ways, and to pass this knowledge on to others (Latour 1994a: 801–803). At one stage, nonhumans pass on their properties to humans; at another, humans "socialize" nonhumans by implanting purpose in them, by interpreting them in ways compatible with the community's requirements for intelligibility, and by extending and limiting their use in communal life.

Political ecology is, according to Latour, a socializing stage. The preceding "technological" stage is characterized by the fusion of science, organization, and industry. Scientists map genomes, fabricate artificial hormones through complicated sequences of chemical transformations, and clone sheep. Agribusiness commercializes these achievements; drug companies and government regulators negotiate their safe applications. All these parties see the material under their purview as just that: matter, not a right-bearing subject. Still, their view of matter has evolved. Matter now behaves like a "complicated organization." It is not merely a passive substance that takes on properties additively. As theorized in systems ecology, complex organizations of elementary units take on characteristics similar to intelligence and purposiveness. Such organizational understanding prepares the way for the next shifting of competence. "Technologies have taught us to manage vast assemblies of nonhumans" such that "our newest sociotechnical hybrid brings what we have learned to bear on the political system" (Latour 1999a: 202). Latour identifies political ecology with the idea of granting some kinds of rights to nonhumans. Nonhumans are "tak[ing] up some of the properties of citizenship" (Latour 1994a: 796–797).

Toward what end are we directing these "assemblies of nonhumans"? What sorts of problems do we face such that the planet needs large-scale management? Latour opens his book *Nous n'avons jamais été modernes* [*We Have Never Been Modern*] by listing a number of peculiar entities that we now encounter every day in the news: the "hole" in the ozone layer, forest fires that threaten endangered species, whales tagged with radio transmitters (Latour 1991: 7–9). The peculiarity of these phenomena is that they are all "hybrids." Hybrids are neither things-in-themselves nor simply social constructions. They are mélanges of something that transcends human control and of actions imputable to mankind. Political ecology is about treating the problematic consequences of hybridization in a "nonmodern" way.

For Latour, two "practical ensembles," ordinarily kept scrupulously apart, form the constitution of modernity. Moderns believe that nature has an independent objective existence—that natural phenomena exist quite apart from any human construction. Scientists set themselves up as nature's spokesmen. Their studies, instruments, and experiments give voice to nature. They represent nature to the rest of us—nature as it really is, undistorted by superstitions and anthropomorphisms. Society, on the other hand, is a human construction. Starting with Hobbes's theory of the social contract, moderns understand society as a pact decided freely among individuals. Social-contract theorists invite us to imagine creating society de novo, as if employing a purified form of reason shorn of ancestral myths and customs. The tendency of modern thought is to undermine ancient notions that political life must take its purposes and limits from nature. Society comes into existence when individuals choose a representative body that gives voice not to nature but to their collective will.

What if the two modes of representation come into conflict? What if nature's scientific representatives make some claim that contradicts the pronouncements of the people's representative body? Moderns, says Latour, essentially deny the possibility of conflict. They recognize two entirely separate modes of representation. Politics is excluded from the laboratory so that scientists may speak for things. At the same time, nature's things have no direct bearing on political representation, which is a matter of will and artifice. Should a controversy arise as to whether to classify something as natural or as artificial, moderns reaffirm their acts of conceptual "purification." They fractionate the phenomenon into its "natural" and cultural components, thus preserving each side of the modern division in its purity.

This divisional arrangement must be understood as a whole, like the separation of complementary powers specified in a political document. Nature and society are separated in the "constitution" of the modern world. Moderns expect the increasing power of science and technology to ease the lot of humankind, and they expect to be heard as equals in a representative political system. Yet this constitution has a hidden function: it facilitates the proliferation of hybrids. While the polity deliberates and legislates over social concerns of every sort, scientists are generally free from external interference.

"The essential point of the modern Constitution," according to Latour (1991: 53), "is to make the mediating activity, which assembles hybrids, invisible, unthinkable, unrepresentable." Scientists go about manipulating matter and living things, setting loose hordes of previously unknown substances and creatures to roam our world. Premoderns sacralized nature, and often they feared that foolhardy human conduct might bring nature's wrath down upon them. Because of these fears, Latour hypothesizes (ibid.: 22), premoderns were attentive to hybrids and inclined to regulate them. But moderns have been so confident in nature's transcendent permanence and its purposeless impersonality that they have allowed the production of hybrids to proceed at a dizzying pace—and, until recently, they have kept them *politically* invisible. Today, however, hybrids have become so prolific that they cannot be ignored. From global warming to the disposal of nuclear waste to genetically altered plants, we find ourselves surrounded by hybrids and anxious about their effects. Yet for Latour there can be no question of becoming anti-modern, of attacking science or renouncing technological change. When he calls us "exchangers . . . of morphisms," he posits that we humans are too interwoven with the objects we engender to make such a project even comprehensible. Moreover, an attack on technological society per se presupposes a totalizing philosophy, tracing all our troubles to some overwhelmingly powerful central cause. Totalizing philosophies invite totalizing solutions, which end up taking the form of political domination and imperialism (ibid.: 168–170).

The problem with hybrids is not that they are generally destructive (although some are). It is that they are *unrepresented*. By making them unthinkable, the modern constitution also exempts them from collective examination, deliberative judgment, and regulation. The challenge for political ecology, as Latour sees it (1991: 194), is to give hybrids seats in our representative assemblies, to "replace the mad proliferation of hybrids by a production that is regulated and decided in common." An assembly in which scientists, industrialists, ecologists, urban planners, and meteorologists debated and legislated regarding hybrids would constitute, Latour proposes, a "parliament of things."

Networks of such groups already exist informally and clandestinely. If one examines whether the Amazon rainforest is giving way to savannah, for example, one encounters a factual controversy shaped not only by botanists and geophysicists but also by human-rights activists, international organizations of ecological activists, and a variety of government agencies (Latour 1994b: 98–103). This network, however, is out of the public eye, and its constituents are never brought together for common deliberation. A parliament of things, in contrast, would effectively give representation to hybrids by creating an arena of negotiation for all the groups whose activities allow us to see the contours of the hybrid. The moral effect of such an institution would come not from its applying a priori ethical schemata to environmental issues but rather from a "slowing down," a "moderation," in the production of hybrids (Latour 1991: 194; 1999b: 177). The very requirement that the production of hybrids be debated—that is, the exposure of scientific and administrative decisions to public deliberation—would itself give us the time to experiment more cautiously with new technologies and environmental practices.

Decentering Ecologism

Theoretical French ecologism often takes issue with the rhetoric of the French ecology movement. Latour (1999b: 42) complains that some ecologists wrongly claim an a priori understanding of the unity of natural phenomena. Others, particularly leftists, try vainly to apply the logic of class struggle to environmental crises (Latour, Schwartz, and Charvolin 1991: 38–41). But Latour, following the usual pattern of French ecologism, never declares himself either an anthropocentrist or an ecocentrist. Still, anyone steeped in English-speaking ecologism will immediately understand that Latour's tolerant view of hybrids precludes his making respect for nature's intrinsic value the touchstone of political ecology. Any attempt to derive norms from nature, whether composed of individuals or of ecosystems, runs up against his charge that "nature" as we understand it is already an artifact of the ethically charged settlement embodied in the "modern constitution." In Latour's pragmatogony as much as in Moscovici's history, there is no "nature" independent of human interests and practices that we might use as a standard for preservation or restoration. As the rainforest case suggested, every "natural" thing appears to us as such only because it has been bundled into a network of knowledge claims, artifacts, instruments, and political influence. It is that web of mediating practices that we should evaluate, not nature itself.

In fact, from Latour's perspective any attempt to defend "wilderness" or the intrinsic value of its inhabitants would appear as a typically modern, typically self-deluding act of "purification." Creating "two entirely distinct, ontological zones" for things human and things nonhuman is the mental habit that keeps us from coming to terms with hybrids (Latour 1991: 21). For Latour's political ecologist, "Nature is not that harmonious and loving mother which we must avoid touching"—the nature that has intrinsic value. The whole point of political ecology, as Latour imagines it (1993: 26), is "to add artifice, indecision, uncertainty, to add scientific to political mediation." Reintegrating artifice means recognizing the process of human-nonhuman interchange for what it is and evaluating it deliberatively.

At times, Latour approaches ecological questions so pragmatically that he will leave many environmentalists uncomfortable. Take his discussion of genetically modified crops. Should genetically altered bacteria designed to transfer genes that make strawberry plants more resistant to the cold be released in field experiments? Latour insists that to proscribe all such work because it involves "unnatural" organisms would be to suppose a priori things that can only be learned a posteriori: whether the organisms do or do not spread beyond a small area, whether they do or do not take on new functions in the local soil ecology, and so on. He recounts how genetically altered bacteria were first produced in an airtight biotechnology laboratory, how a "suicide" gene was added to them, how soil biologists worked to follow the bacteria once they were released in field experiments, and so forth. Objecting to such procedures out of fear that bacteria might spread uncontrollably is to *presuppose* that there is *an* environment. Yet, says Latour, whether phenomena are interconnected can only be a matter for empirical investigation. We are engaged in "collective and permanent experimentation" in respect to our surroundings. The task of the human sciences is not to forbid or limit this experimentation; it is to "transform partial experimentation into complete experimentation, to prevent its being limited to a segment of the collective (scientific researchers) and to a single moment (establishing truth)" (Latour 1991: 42–43).

Is Latour's experimental ecologism anything more than a convoluted anthropocentrism? Latour thinks that an institution like a parliament of things would "transform partial experimentation into complete experimentation" (1991: 42). What does a parliament of things do if not bring a wider assortment of *human* interests (wider than those of scientific researchers) to bear on experiments? On what grounds could representatives debate about the use of genetically altered strawberries? We know that they would not be worrying about violating the nature of the strawberries themselves. So wouldn't they have to be weighing various risks and benefits to human welfare (e.g., the danger of doing unintended damage to other valuable crops) against, say, consumers' desires for succulent fruit in wintertime? If this parliament assembles representatives of business and of the state, ecology groups, workers, etc.—all those who constitute the networks that already clandestinely track hybrids—what will they represent other than their own understandings of the requirements of *human* well-being? In the final analysis, things do not represent themselves. As one of Latour's critics insists, "even in this extended parliament it is only people who do the talking" (Harbers 1995: 274). What Latour proposes is really a parliament *about* things, not a parliament *of* things. Thus, perhaps he is a species of anthropocentrist after all. But Latour refuses to think of himself that way. He insists that his theory of the exchange of properties among things in a collective means that we go "from being anthropocentric to becoming *decentered*" (Latour 1991: 54).

Comparing Latour's ideas with those of an English-speaking ecologist who also analyses the politics of technology once again highlights the patterns—and the limitations—of each community's environmental argumentation. In *The Whale and the Reactor*, Langdon Winner (1986: 19) investigates how "artifacts have politics"—how "they can embody specific forms of power and authority." Just as Latour's pragmatogony observes "crossovers" between human and nonhuman orders, Winner argues that the politics of artifacts becomes *our* politics. Technologies are not just passive instruments in our hands; they change our behavior in ways that deserve more critical scrutiny than we ordinarily allow. Too often they make us subservient to authority, deprive us of meaningful work, or subject us to new forms of psychological manipulation. Without proposing anything quite like a parliament of things, Winner wishes for a more deliberative assessment of technoscientific choices.

Closer analysis of Winner's argument, however, shows how ideas similar to Latour's, when assembled in a rhetorical field polarized by centered ecologisms, get driven toward anthropocentric and social constructivist conclusions. Winner feels moved to measure his sociotechnical view explicitly against assertions of nature's intrinsic value. This drives him into an analysis of the meaning of "nature." He finds that among English-speaking writers ecocentric understandings of nature coexist with instrumentalist-economic and Hobbesian-survivalist understandings. "Nature," he surmises (1986: 131–137), is a "social category" open to "myriad interpretations." Rejecting every essentialist interpretation of "nature" appears to incline English-speaking theorists toward social constructivism. But then, in backing away from any particular interpretation of nature, and particularly from an ecocentrist one, Winner lands squarely in anthropocentric territory.

Winner's version of a more deliberative ecologism suggests that, at the "genesis of each new technology," the "political wisdom of democracy" be brought to bear upon it. That rarely happens today, because certain people seek "dominion over others" through technologies. If certain technologies have disempowering effects, this is because certain human agents managed to inscribe their interests into their design. Winner wants to shift interest-defining prerogatives away from them and to a wider public. Then we could make decisions about technologies in the context of a much broader question: "What forms of technology are compatible with the kind of society we want to build?" (ibid.: 50–58). That question extends the range of concern beyond issues of efficiency and social control to aspirations for social justice, worries about the abuse of power, desires for more decentralized authority structures, and interests in human health. These are all anthropocentric interests. From the diagnosis of the problem to its remedy, Winner takes for granted the existence of centered agents. An elite that designs technologies to distribute social goods to its own advantage, just like the public that makes technology accountable to its interests, consists of human agents who are the locus out of which decisions flow. These agents either already know how to manipulate their world intentionally or can be put in a position to learn how to do so. They and their world are at least potentially distinct from each other. Winner supposes that purification is possible.

Latour's principle of symmetrical observation forces him to study not only the social construction of nature but also, and equally, the natural construction of society. "Yes, society is constructed," he admits; "but not just socially constructed"—because "society" has acquired its form by swapping properties with nonhumans, just as their properties develop from being "socialized" (Latour 1994a: 793; 1999a: 198). In Latour's pragmatogony, the tasks of political ecology arise from earlier understandings of the properties of nonhuman systems. Yes, there are human agents who now talk of granting rights to nonhumans in order to oppose the accelerating production of hybrids. Their opposition is not simply an act of will, a populist appropriation of political initiative from self-centered elites. It presupposes, Latour says, agents whose very ability to conceive of environmental protection in this way stems from lessons learned through "technoscience" (1999a: 203–204). The technoscientific coordination of research, organization, and industry (e.g. the genome-sequencing project) elicits unexpectedly complex behavioral properties in matter. It is this complexity that makes nonhuman phenomena eligible to become the focus of new forms of environmental concern.

At the same time, Latour calls for careful attention to the construction of political terms too. Interpreting Hobbes, for example, one witnesses how concepts of "power" and "representation" get defined in full awareness of the need to secure political authority against potential challenges made in the name of truths established through independent empirical experimentation (Latour 1991: 42–43). Symmetrically, one assumes neither that society is the ground of science nor that scientific understandings of nature are the ground of politics. An ecologism that is neither anthropocentric nor ecocentric requires painstaking, empirical, "anthropological" fieldwork that maps out in detail the networks in which exchanges between these domains takes place. Its political effect comes from turning informal, clandestine exchanges into overt, socially negotiated ones.

This is not to give noncentered ecologism the final advantage. The lessons of cross-cultural theoretical comparisons flow in both directions. Winner's perspective exposes a missing side of the edifice of Latour's political ecology. Winner's political agents know their interests and take advantage of power disparities in society to satisfy them disproportionately. In comparison, Latour's hybrids seem blankly disinterested. Latour focuses so much on making us aware of hybrids that he virtually disregards how the public's view of them may be limited by interested behavior backed up by differential power. To use the example of genetically modified organisms (e.g., freeze-resistant strawberries): Latour gives no hint that what is known of the behavioral properties of GMOs depends on minimally regulated research conducted by profit-driven agribusinesses. There is something too dispassionate in Latour's description of humans' and nonhumans' switching characteristics. Winner keeps us alert to how power differentials may bias that entire process.

Latour's suppression of the concept of interests also makes his proposal for a parliament of things baffling at two levels. Whatever he means by 'representation' in this context, it is clear that he cannot be accused of wanting people to represent the interests or the well-being of nonhumans. Representation in a parliament of things really means designing a deliberative process in which not just competing human values but also all the effects and uncertainties of technological innovation and empirical research get discussed. But Latour gives no account of how different types of claims made in this parliamentary body could be moved discursively toward resolution.

Suppose that the entire collective surrounding genetically modified strawberries—botanists, agronomists, geneticists, consumers, environmental activists, agricultural unions, and so on—meet in a single chamber. Now what? Who says what? How does a claim that strawberries may pass their genes to related plant species interact with a dispute about statistical models of pollen distribution? Then, how does that ungainly debate mesh with worries about whether farm workers who bring in late-season harvests should get special compensation? Missing from Latour's concept of representation is any analysis of the discursive conditions of consensus.

Perhaps the parliament of things is not supposed to reach a consensus at all. Perhaps all it needs is a decision rule through which it can close debate. But this interpretation only displaces the source of bafflement. In a legislature, decisions get sealed by votes. Who has the right to vote and how much weight each vote is allotted must be established before balloting takes place. It is utterly unclear how such commonplaces of legislative practice could be approximated in a parliament of things.

There is no limit to the number of "actants" with a claim to be represented on an environmental issue. To debate issues having to do with the depletion of atmospheric ozone, Latour's parliament would include representatives of the chemical industry and their workers, a meteorologist, and a spokesman for the State. Does each of them get one vote? Are votes distributed in proportion to the number of people they represent, or in proportion to something else? In any case, why stop with such a limited legislative roster when the "network" that politicized this issue certainly does not? Let's include consumers who want aerosol products and oncologists who represent skin-cancer patients. Let's represent each suspect chemical with biochemists, some of whom argue that their compound is benign and others who regard it as harmful. Let's include marine biologists who fear that increased ultraviolet radiation will kill plankton. Let's include activists for the Third World, whose people may need the cheapest refrigerants available. There is no end to this list. Yet the outcome of any parliamentary vote obviously depends on the composition of the parliament.

At times, Latour further embroiders his parliamentary scenario with an argument for bicameralism. He imagines an "upper house" that decides the number of actants involved in a controversy and a "lower house" that deliberates over whether such actants can be accommodated in the good life of the community (1999b: 157). Disappointingly, none of this filigree stabilizes the composition of a parliament of things in a way that makes it possible to count votes in a consistent fashion.

Postmodern Ecologism?

Contributors to the politicizing stream of French ecologism negotiate the nature/culture divide differently than systems theorists. Although advocates of a systems approach become reflexively sensitive to its historicity and to its potentially authoritarian interpretations, they nonetheless believe that it represents a *progressive* moment in human understanding. The epistemologies of Latour and Serres make it impossible to conceive of any science, including ecological systems theory, as a moment in the progressive, dialectical disclosure of truth. For Serres, "to think is to connect and disconnect things in circulation, to cross the transcendental space of communication in every direction, to intercept and exchange the forms of that space" (Crahay 1988: 74). Latour insists that his investigations of science and society are not dialectical. There is no tale of transcending contradiction by the growth of an ever more comprehensive synthesis, no odyssey in which humanity comes gradually to self-knowledge.

The project of politicizing nature sounds like a "postmodern" project. According to Jean-François Lyotard (1979: 63), "grand narrative has lost its credibility" in a "postmodern" culture. Modernity's master narratives included a story of human progress through the domination of nature, its objectification, and its subjection to technological control. In the process, science dissipates irrational custom and myth. Removing the veils of irrationality would be the path to enlightenment. Moderns recount history as the unfolding of universal rationality. Postmoderns regard each component of this narrative as a construction ripe for deconstruction. Human subjects are not autonomous moral agents; they are sites shaped by power, consciousnesses produced by networks of forces (Conley 1997: 4). In Michel Foucault's view, for example, human sexuality is culturally constructed by means of repressive social standards regarding care of the body and acceptable forms of behavior. Nature is not objectively there. In fact, "naturalizing" something, interpreting it as the expression of some unchangeable essence that is its "nature," is one of the most powerful ways of enforcing one normative construction against possible competitors. Modernist Enlightenment presupposes commensurable values so that movement from one conception of well-being to another can be evaluated as progressive or regressive. For postmoderns, the incommensurability of moral discourses is a corollary of the indeterminacy of meaning. In the place of theoretical criticism based on supposedly universal standards or practical criticism embodied in a supposedly universal class, postmoderns typically propose multiple forms of localized resistance against the forces of normalization.

French postmoderns have penetrated ecological political theory—but, curiously enough, more in its English-speaking than in its French variant.[10] I want to suggest that even postmodernism plays itself out differently in French than in English-speaking ecologism, precisely because of differences in the rhetorical fields that I have been emphasizing throughout this book. Postmodernism offers a tempting critical framework in which to oppose ecological theorists who allude to "nature" and "wilderness" as if those concepts were unproblematic. It is much less alluring in a linguistic community in which it is routine to reciprocally problematize nature and humanity.

The motives to apply postmodern thought to English-speaking ecologism are evident in the works of Tim Lukes and Peter van Wyck, each of whom begins with an assault on those most resolute nonanthropocentrists, the deep ecologists. Van Wyck deconstructs the claims of deep ecologists to have overcome the nature/culture dualism. Placing humanity in nature, argues van Wyck (1997: 53), really amounts to a theoretically untenable "move to the outside." Like the most ambitious systems theorists, deep ecologists adopt a perspective that supposedly embraces the whole earth, turning it into an organic totality of which humanity is a part. Most troubling for van Wyck are the effects of this totalizing move on understanding the self. The deep ecologist proposes overcoming humanity's separation from nature by making the self merge with the organic wholeness around it. Van Wyck (1997: 75–102) objects that this new universal self overrides the actual heterogeneity of human subjects. Even more unexpected, "deep ecology projects a covert anthropocentrism" (ibid.: 100). When nonanthropocentrists praise the traditions of primal societies because they do not break sharply with nature, they actually "move the primitive directly into the field of desire" (ibid.: 98). At the same time, deep ecologists long for a wilderness so free of human contamination that they effectively erase aboriginal peoples from existence.

Lukes likewise detects "soft anthropocentrism" in deep ecology. Nonanthropocentrists "construct Nature as an active subject" while in fact seeking the realization of their own subjectivity in it (Lukes 1997: 11). In effect, their stance "favors the nature-regarding interests of humans" (ibid.: 27). But more anthropocentric environmentalists fare no better when scrutinized through Lukes's quasi-Foucauldian lens. Environmental groups of all varieties employ "discursive frames" regarding "nature" to legitimize a vision of how power should be deployed. The Worldwatch Institute, for instance, reduces "Nature . . . to a system of systems" that can be monitored and redesigned. Its program of sustainable development foreshadows the extension of "disciplinary projects" as the complement to its "global surveillance sweep" (ibid.: 91–94).

In France, a similar line of postmodern analysis is not unknown. But the absence of theorists advocating strongly centered ecologisms diminishes the force of the critique. Postmodernism comes across less as a bold corrective to nonanthropocentrist naturalism than as a skeptical plea for ecologists not to confuse well-intentioned rhetoric and legislative goals with actual accomplishments of environmental policy. Pierre Lascoumes's *Écopouvoir* [*Eco-Power*] shows what happens when a sociologist influenced by postmodern thought strikes out across the tangled field of ecological discourses without encountering a deep ecologist.

Lascoumes maintains that, since the 1970s, French environmentalism has evolved from a popular movement grounded in a critique of technological society to a technocratically dominated policy area. Displacing democratic mobilization, scientific experts, engineers, and technicians now form an "eco-power" that sees to the "rational governance of the living milieu" (Lascoumes 1994: 32–33). Lascoumes warns that we may be witnessing the accession of a dangerous new form of repressive authority, a second stage in the constitution of "bio-power." Michel Foucault (1976: 183 ff.) described the first stage as the result of a dramatic change in the exercise of sovereignty that took place in the nineteenth century. Social conformity would henceforth be enforced less by disciplining individual bodies than by monitoring, regulating, "normalizing" the processes through which entire social groups develop their identities. Eco-power inaugurates a second stage, hypothesizes Lascoumes, by extending surveillance and control beyond human society to every other form of life. Ecology then "transcends its sectoral project and acquires the ambition of guiding political choices towards a general optimization of the world" (Lascoumes 1994: 313). Every new regulation for public hygiene, every law concerning pesticides or agricultural runoff involves issues of biology and epidemiology far beyond the comprehension of the general public. The development and application of such rules falls more and more into the hands of agronomists, engineers, and other experts. Lascoumes' thesis is that an ecological discourse framed in terms of "scientific and technical rationality" is what legitimizes eco-power and "secures a situation in which its experts have uncontested mastery" (ibid.: 28).

In a Foucauldian framework, discourses of all types—medical, penal, psychiatric—combined with attendant social practices, *constitute* individuals in ways that fit them for domination by various social groups. Behind such domination, there is no truth to be discovered, no human essence to be liberated. We must study social relations, argues Foucault, "in accordance with the intelligibility of struggles, of strategies, and tactics" (Foucault 1980: 114). Struggle is all there is. Lascoumes appears to have accepted much of Foucault's perspective. Lascoumes observes, first of all, that there is no "environment" (Lascoumes 1994: 9–14; 1991: 61–65). There is no substantive referent which all who express concern for "*the* environment" have in mind. Farmers and industrialists, rural life associations and birdwatchers all try to problematize different aspects of their bio-physical surroundings. It is meaningless to say that concern for "*the* environment" motivates anyone's action. Instead, various groups *constitute* "the environment" by highlighting situations of concern to them. Second, the divergent interests of groups can only be adjusted—not reconciled—because there is no common measure for weighing interests (ibid.: 104–105). Through a process that Lascoumes calls "transcoding" (ibid.: 22–23), groups present information so as to create the "cognitive framework" in which later decisions about private and public action will be made. Because there is no such thing as "the environment," Lascoumes holds that those who talk of protecting "nature" can only mean protecting *their* power and interests against those who would "code" issues differently. Even ecologists who would "make nature a subject of law" are really only "self-validating their own position as the legitimate representative of trees and streams" (ibid.: 10). Competing *social* interests valorize things that count as "the environment." "Nature" is an effect of their transcodings.

What is the political lesson of this argument? In postmodern fashion, Lascoumes (ibid.: 27–31, 263–264, 297) wishes for the development of a green "counter-power" to resist the growth of official eco-power And much like Latour, Lascoumes (ibid.: 287) favors the invention of "hybrid forums" of democratic participation. To prevent the accumulation of unchecked power in the hands of scientific and technical experts, new arenas (e.g., ecological hearings and citizen conferences) must be constructed in which representatives of the widest range of interests, dissident scientists and members of the general public exercise genuine influence on policy making. Following *Écopouvoir*, Lascoumes (1998, 1999) has, with some optimism, explored the precedents and prospects for these alternative arenas.

At two levels, however, Lascoumes's ideas are in tension with their Foucauldian framework. First, tying knowledge and power so closely together prevents Lascoumes from coming to any distinctively ecological conclusions, despite his obvious sympathy for green causes. Thus, when he talks about ecologists as a "counter-power," it is more their role in resisting the accumulation of power per se than in championing any "environmental" issue such as protesting the contamination of underground aquifers or the loss of biodiversity that he admires. Since in his view there is no such thing as the environment, he can only call a policy "environmental" because other people have done so in order to express their utilitarian, hygienic, aesthetic, or spiritual preferences. Without a conception of "nature" or the "environment" that gives the world some standing outside of social codes, Lascoumes is sometimes reduced to describing "environmental" policy in pluralist fashion, as whatever measures result from the adjustment of interests among competing groups. (See, e.g., Lascoumes 1994: 107, 271.) Second, for the moment, at least, the theme of eco-power blows the dangers of technical expertise out of proportion in relation to more commonly acknowledged problems in environmental decision making. Lascoumes has written one of the most fascinating and detailed books of case studies of environmental policy in France. The lesson of those cases, however, is arguably quite different from the one he draws. They illustrate less the unchecked authority of scientists and technicians than the *weakness* of the French Ministry of the Environment and of environmental civic associations, including their expert advisers. Often Lascoumes's evidence shows how interest groups favoring the rapid expansion of industry and infrastructure—real estate developers, manufacturers, agribusinesses—have played a preponderant role in shaping environmental legislation and its implementation. Such evidence suggests that what distorts public deliberations on environmental policy is not only the exclusion of various non-expert actors from the political process. It is also—largely—the influence of monied interests in determining the community's priorities. In that case, better environmental protection depends more directly on the emancipatory "modern" project of devising political forms in which democratic norms of equality and undistorted communication allow the full range of community interests to be represented than on the "postmodern" project of exposing every legitimizing discourse as a form of power.

Postmodern critique is useful when it uncovers politically dangerous extensions of centered ecologisms. Like Edgar Morin, René Passet, and many other French ecologists, Lascoumes calls for vigilance against the potential for ecologism to turn authoritarian if it takes its cues only from scientists and technicians. To redeem his ideas as potential contributions to social criticism, however, two amendments are necessary. First, it must be the case that "the environment" or "nature" stands for something more than a series of codes employed by rival social groups to incline policy in their favor. Conceding that "nature" is not identical with human interests in the world need not lead back to a positivistic epistemology. Latour's anthropological investigations into the complex networks of humans and nonhumans that constitute scientific "facts" present a significant alternative. Lascoumes's own predilection for hybrid forums suggests a similar option—provided that we can view the forums as opportunities to elicit a *better* understanding of the world (e.g., to bring out factual observations and ethical considerations that experts might ignore), not merely as another arena for social competition. Second, Lascoumes's amended theory should not rule out the possibility that people who become more aware of how social codes and hidden power operate can use that knowledge to develop strategies to overcome these obstacles. Lascoumes's empirical case studies could then contribute to developing a more ecologically rational social project.

Postmodern critique becomes seriously counterproductive, however, when a fascination with incommensurable discourses takes the place of any attempt to grapple empirically with a world undergoing rapid ecological deterioration. One last and particularly egregious example shows how far wrong postmodernism can go.

In his 1994 book *The Illusion of the End*, Jean Baudrillard lambastes Serres's book *The Natural Contract*. In a chapter titled "Maleficent Ecology," Baudrillard (1994: 78–88) asserts that the central problem today is that "the human race is beginning to produce itself as waste-product, to carry out this work of waste disposal on itself." We do this, he charges, not only by polluting but also by building, by paving, by advertising, by turning events into news, and by laying off workers. It is a "symbolic rule" that a rising view of "human-beings-as-waste" is "accompanied by a new human rights offensive." Rights are what we invoke when the right-protected quality has already disappeared. On this reading, Serres's "natural contract amounts to a definitive recognition of nature as waste." Baudrillard continues: "We have objectified [nature] to death, and this ecological cover merely asserts our right to go on doing so." By treating nature as a right-bearing subject, we discount its radical otherness, especially its "maleficence." We are on the path to Biosphere 2—to environmentalism as the "optimal management of residues."

It is hard to tell whether Baudrillard read beyond the title of Serres's work. He evinces not the slightest awareness of Serres's inquiry into the relationship among nature, contracts, law, and trials. For Baudrillard, to judge the merit of ecologists' prescriptions it suffices to shroud them in disturbing images: managed residues, charnel-houses, comatose nature preserved in a glass coffin. And what lesson is to be gained from this sophistical word play? "It is the inseparability of good and evil which constitutes our true equilibrium. . . . We must not reconcile ourselves with nature." Does this mean that we should refuse immunization against diseases produced by nature's "maleficence" or tolerate drought by going hungry? Who knows? Baudrillard presents us with the case of a writer who ostensibly has something to say about ecology—about the fate of a world undergoing climate change, rapid loss of animal and plant species, and deforestation and desertification in some areas—yet who in fact does little beside posit "symbolic rules" that give him a pretext for unleashing a torrent of cynically Nietzschean bons mots.

Many major contributors to French ecologism are either innocent of postmodern inclinations or positively hostile to them. André Gorz (1988: 157) belittles structuralism and its postmodern descendants as "the ideology of triumphant technicism" precisely because their critique of all rationality makes individual autonomy meaningless. For Edgar Morin (1993: 88), "the critique of modernity . . . has given birth to an impoverished postmodernism which only consecrates an inability to imagine a future."

This trenchant comment about his contemporaries by Serres's (1992c: 24) helps to explain the tension between postmodernism and French ecologism:

Many of today's intellectuals are the direct inheritors of the preceding generation, which was trained in the humanities and human sciences. . . . We must praise this generation, of course, but adding that it was totally ignorant of another side of cultural modernity . . . : mathematics, thermodynamics, physics, chemistry, genetics, and so forth. As a result, contemporary political reflection is often conducted by people whose education I would call semi-paralyzed.

Indeed, Latour, in an ingenious turn of argument, contends that ignorance of contemporary science gives some postmoderns unsuspected affinities with moderns (1991: 84–87). Postmodern "hyper-incommensurability," says Latour, is only a new guise for free, critical subjectivity—the counterpart of objective nature in the modern constitution. Moderns from Descartes to Kant to Sartre have asserted the powers of the human subject to reason and criticize independent of "natural" determinations. "Postmodern" denunciations of the centered human subject do not so much deconstruct this freedom as exercise it with a vengeance. Each generation, it seems, launches a more "dizzying succession of critiques" than its predecessors. The lesson of Baudrillard's interpretive promiscuity is that endless critique, in the absence of actual investigations into contemporary practices and disciplines that present the nonhuman world to us, degenerates into little more than a facile assertion of *human* will. In fact, it becomes anthropocentrism of the most self-indulgent sort.

Noncentered ecologists keep their theories from spinning off into a speculative void by tethering their ideas to certain preferred methods of quasi-empirical study: Moscovici's historical inquiries into relationships of technologies and work patterns in various "states of nature," Morin's dialectical "anthropolitics." Along similar lines, Serres and Latour propose symmetrical studies of nature and society. To chart "political" relations that scramble boundaries between nonhumans and humans, they propose a research program that demands far more than parsing language structures and deconstructing social codes. Their research takes them into laboratories. It involves reading scientific journals, looking at the political skills of scientists and engineers, analyzing scientific controversies and trials as moments when "nature" gets constituted, and engaging in anthropological studies of human-nonhuman interaction. Although Serres and Latour treat science symmetrically, both men understand that their epistemologies give them no license to regard science as a symbolic system whose logic might be deconstructed with a few tools borrowed from semiotics and psychoanalysis, without delving into how scientists make, support, and disseminate their claims. French critics—Luc Ferry as much as Jean Baudrillard—miss this reflexive realism.

Reflexivity of this sort renews the heritage of humanism. Traditionally, humanism implies the cultivation of taste and a disposition to seek understanding of humanity's aesthetic and moral achievements (Levi 1969: 16). Humanists—the caretakers of philosophical, literary, and artistic learning transmitted to the present from millennia past—preserve and cultivate this vast legacy of ways of valuing the world. Observers who label French ecologism "humanistic" do so not because they confuse "humanism" with anthropocentrism but because they see connections between ecological thought and this rich heritage (Sandoval 1995: 116–117; Chalanset 1997: 5, 13–16). Serres's erudite juxtapositions of philology, the history of law, and ancient philosophy are evidence of this. In his view (1992a: 259, 266), ecologism depends as much on the human sciences (because they teach tolerance by undermining old prejudices) and on the humanities more generally (because they investigate rigorous knowledge in myth and literature) as on the natural sciences.

But humanism must be renewed if it is not to succumb to "semi-paralysis." Latour (1991: 186) calls for a "redistributed humanism." A redistributed humanism prizes our dual role as the source of quasi-objects and the recipient of their properties. Latour sees the possibility of "Enlightenment without modernity"—that is, a continuing quest for scientific and social understanding, absent the modern bifurcation of nature and culture and shorn of modern habits of ideological denunciation. Renewing humanism requires bringing "cultural modernity," as Serres puts it, into closer contact with contemporary science. His appointment to the Conseil pour les droits des générations futures [Council for the Rights of Future Generations], like Morin's service as president of the Conseil scientifique de la consultation natoinale sur les lycées [Scientific Council for the National Consultation on High Schools] and Jöel de Rosnay's work of popularizing ecological understanding at the museum known as the Cité des sciences et de l'industrie, indicates that a humanistic ecologism has among its primary targets the educational practices of the institutions through which society produces citizens who will exploit or care for the earth.

The lesson of nature's politicizers is that contact between the sciences and the humanities entails far more than giving technicians and humanists the missing complement of their own education. Ecological reflexivity means becoming conscious of the interpenetration of different branches of knowledge. For Serres and Latour, only deeper awareness of the founding concepts that lead us to engage destructively with the world could allow us to imagine ways of mitigating the undesired effects of our own activities.

5

Personalizing Nature

Those who politicize nature argue that both human society and nature are structured by cross-cutting *collective* understandings of law, contract, and representation. Where, however, is the *person* in the understandings of ecologism that follow from these theories? What becomes of the subject who perceives, who chooses, who intends, who loves? Does the project of politicizing nature come at the cost of dissolving the individual consciousness in a stream of collective representations, reducing the person to a being who simply perpetuates culturally transmitted conceptions of power and right? Is there a danger that certain conceptions of nature may negate the individual's status as a creative actor or as a spiritual being? If respecting the unique identities of "natural" things requires limiting human prerogatives, is the concept of free individuality an obstacle to ecologism—or, in some sense, its precondition?

In France these questions arise most frequently in a personalist perspective. Personalists place individual *épanouissement* [self-realization or flourishing] at the top of the ecological agenda. In this regard, their perspective bears a certain resemblance to the "ecological virtue" approaches proposed by some English-speaking ecologists in the 1990s (Barry 1999b; O'Neill 1993). But the long pedigree of French personalism gives its French proponents a different cache of ideas to draw upon. In 1903 the founder of French personalism, Charles Renouvier, wrote: "The term 'personalism' assigns us the task of demonstrating, with reasons—at first logical, then moral—that knowledge of the person as consciousness and will is the foundation of all forms of human knowledge."[1] Renouvier meant to assert the primacy of the personality and its spiritual quest over any positivistic analysis of human behavior. No biological or social conditioning could give an adequate account of the moral decisions and longings for transcendence at work in every individual. Bordering on idealism, personalism locates the center of awareness in the personality. Anticipating existentialism, it insists on the free choice through which the individual seeks out moral engagement. Values become manifest in everyday life as people express love and negotiate tensions in a spirit of accommodation and mutual respect. Personalists are sometimes—not always—religiously inclined.

In the 1930s, the Catholic intellectuals Emmanuel Mounier and Denis de Rougement brought personalism to its political apogee. The journalism of their "nonconformist movements" challenged the spiritual impoverishment of a materialistic society. The influence of personalism extended into the formative years of political ecology thanks to Bernard Charbonneau and to Denis de Rougement himself. They pioneered the application of personalist ideas to environmental matters by arguing for the complementarity of a flourishing sense of personal freedom and a flourishing "nature." Today, Jean-Marie Pelt and Philippe Saint Marc continue this legacy. Rather than construct a formal environmental ethics out of interrelated definitions, principles, and rules, they write essays whose moral effects depend on appealing to aspects of personhood that are suppressed in a materialist civilization.

Just as Mounier's thought can be called a philosophy of "ambivalence" (Pierce 1966: 50) because of the multiple, even contradictory, values it embraces, so internal tensions run through personalist ecologism as a whole. Rougement and Charbonneau are critics of state power and so advocate a radically decentralized, federalist ecologism. Saint Marc leans in another direction entirely, toward a state-led "socialization of nature." Like Rougement, Saint Marc and Pelt believe that the spiritual dimension of personalism connects with a religious faith in a morally grounding Creator. Charbonneau remains agnostic. Pelt is heavily indebted to scientific systems ecology; the spiritual preoccupations of other personalists make them suspicious of "materialist" science and lead them to prefer appreciating "nature" sentimentally.

At times personalist humanism verges on anthropocentrism. *Épanouissement* can designate a form of moral perfectionism in which an appreciation of nature has a place alongside more conventionally recognized virtues as a condition of human self-realization. In that case, the most one can say about the French difference is that, unlike their English-speaking counterparts, Pelt and Saint Marc leave their anthropocentrism unavowed and never feel constrained to shape their views to defeat ecocentrism. Other French thinkers, however, set forth a view of *la personne* that is not perfectionist and that leaves ecologism noncentered. They portray a self whose psychic fractures engage with the world in ways that give rise to multiple conceptions of a nature worth protecting. This line of theorizing, anticipated by Charbonneau, is developed most suggestively by Denis Duclos in works that reinvigorate ideas commonly associated with personalism.

Personalism and the State: Regionalist Ecologism

In France, personalism first gathered political influence among "nonconformist groups" in the 1930s. The historian Jean-Louis Loubet del Bayle (1987: 24) explains that "a feature common to all [these] groups was the feeling that what they were living through was a *crisis of civilization*, that is a *total* crisis, which brought into question *every* aspect of human existence." The nonconformists were highly critical of liberal society, of the "regime of parties," and of the "reign of money." Philosophically, they rejected modernity's penchant for rationalism and materialist productivism, both of which neglect the individual's quest for spiritual fulfillment. They declared themselves "revolutionary" not to challenge the state with arms but to call for a radical transformation in *personal* ethics. Emmanuel Mounier's emphasis on the responsible, engaged individual distanced the nonconformists from temptations to affiliate with either of the era's radical collectivist positions, communism and fascism. The very term "person," in fact, was intended to indicate a "third way" that was neither idealistic nor materialistic and which accepted neither statism nor anarchism, neither individualism nor collectivism.[2]

Two men who would end up taking environmental concern into their reflections developed their early political thought while participating in personalist circles in the 1930s. Bernard Charbonneau's 1937 essay on "The Feeling of Nature" is remembered as the "first ecological reflection published in France" ("Décès: Bernard Charbonneau," *Le Monde*, May 2, 1996). "Man," wrote Charbonneau, "returns to nature to find again his condition as man, to run risks, to feel hunger and thirst. . . . He knows that only there will he recover his candor and his lost freedom." To this end, he organized camping retreats in the Pyrenees and in the plains of southwest France so that individuals might shed an artificial life devoted to material pleasures. He envisioned a personalism oriented toward an "ascetic city." Declaring that nature "must be to personalism what class consciousness is to socialism," Charbonneau first developed an argument that the condition of the earth has to be understood in relation to the conditions of human self-realization.[3] With his friend Jacques Ellul,[4] he formulated "Gascon personalism," questioning the moral neutrality of technology.[5] Ellul and Charbonneau believed that technology—not just mass politics—created pressures for social conformity and for alienated human relations.

Over the course of some 20 years after World War II, Charbonneau composed *Le Système et le chaos: Critique du développement exponentielle* (1973). In that work he ferociously attacks state power. He alleges that the state operates less as a corrective to a private sector that sometimes unfairly pursues private gain than as an accelerator of economic growth, and that it subjects individuals to homogenizing methods of administration and economic development, fostering wars and overexploiting nature. Charbonneau admits that technical and scientific progress are not in themselves evil, insofar as they "allowed man to be liberated from the suffering and terrors of nature" (ibid.: 157). But their very success in reducing nature's terrors created the conditions for a world in which "nature" will have disappeared. "If the accelerated growth of a population depending on accelerating production continues . . . , a moment will come when not iron and cars are lacking, but elements: water, space, time." (ibid.: 14)

Similar fears informed the works of Denis de Rougement, but not before the 1970s. A Swiss intellectual, Rougement had begun making a reputation as a personalist some 40 years earlier.[6] In *Politique de la personne* (1934) he maintained that our lives must combine respect for our own freedom as individuals who make choices, create, and take risks with a sense of responsibility toward others as persons and as members of a community (Saint-Ouen 1995: 16–20). His most celebrated work, *L'Amour et l'Occident* (1939), explores the latter theme. The Western conception of love, he argues, encourages a union that nonetheless preserves the distinct identities of its constituent persons. European culture also contains, however, another conception of passionate love—one that aims at the fusion of individuals in a social whole. Well before Germany invaded Poland, Rougement warned that state centralization and nationalism feed on this second conception. The state, he charges, is an instrument for crushing local liberties and for standardizing human conduct. Whether to extend its territory or to protect itself from others' expansionist designs, the state uses war as a pretext for the centralization of power and the regimentation of citizens (Rougement 1977: 93–110).

Rougement extended this critique in an environmental direction after reading *The Limits of Growth*.[7] In *Les Dirigeants et les finalités de la société occidentale* he observed: "Henceforth we are able to calculate the exhaustion not only of habitable space, but of existing oil reserves, forests, uranium, air and water on earth. Everywhere these calculations reveal to us the limits of natural resources." (Rougement 1972: 17) Modern societies' appetite for limitless growth and their penchant for concentrating political and economic power to achieve it were destroying the grounds of civilized life. These convictions led Rougement to found and preside over Ecoropa, a transnational organization promoting a Europe-wide, regionalist approach to ecologism. Other founding members included Brice Lalonde, Jean-Marie Pelt, Serge Moscovici, and Bernard Charbonneau.

Rougement and Charbonneau share a conviction that ecologism adds urgency to personalist arguments for radically decentralizing power. To check the crisis of a civilization dedicated to economic growth and state power, Rougement insists (1977: 218), "we must begin by restoring small units of participation." Centralized European states have often forced localities to accept environmentally destructive patterns of infrastructural development. Rougement counterproposes that "a regional entity . . . has better chances of becoming a factor of regulation and harmonization of growth" (ibid.: 278). Effective responses to border-crossing problems such as air pollution and water pollution are often blocked by states protecting their national interests. Thus, it is also necessary to seek solutions to such problems on a supranational scale—for example, through solutions crafted by all inhabitants of a river basin or even by a Europe-wide federation (ibid.: 280). Charbonneau is more suspicious than Rougement of a supranational Europe; however, he too calls for political decentralization. Charbonneau (1991b: 172) favors a federation "made of a complex of societies, of enterprises more deeply rooted in nature and local reality because they are small." This program may put English-speaking readers in mind of familiar arguments for bioregionalist ecologism. Because of differences in the rhetorical handling of "nature" in France and the English-speaking world, however, comparisons must approached cautiously. The French speak of regions, English-speaking ecologists more often of "bioregions." The prefix, or its absence, is significant.

Bioregionalists, says Peter Berg (1991: 7), take a "biocentric viewpoint;" they argue that "human society is ultimately based on interdependence with other forms of life." Their quest for regional rootedness stems from a conviction that industrial-urban social forms dangerously cut their inhabitants off from life-sustaining ecosystemic processes. Anomie and environmental destruction go hand in hand. One of the most noted bioregionalist theorists, Kirkpatrick Sale, contends that if people lived in small, bioregionally defined communities they would be able to reconnect with the land. They could come to understand and appreciate water flows, soil characteristics, and natural distributions of plant and animal species. Understanding does not imply abstaining from using nature; rather, it means "using all the biotic and geological resources to their fullest, constrained only by the logic of necessity and the principles of ecology" (Sale 1991: 46). In such a formulation, the "bios" of the bioregion is a center of valuation.

In English-speaking ecologism, talk of deference to "the principles of ecology" and of biocentric respect for nature does not go unchallenged. Critics point out that defining features of bioregions—e.g., watersheds or the range of various plant and animal species—often coincide only imperfectly. As a consequence, the appropriate level of aggregation to use in constituting a bioregion is a matter of human choice, not just a matter of the "principles of ecology" (Alexander 1990: 167–169). Human judgment becomes all the more important when it is admitted that bioregional boundaries depend in part on culturally grounded notions of the scale and activities of a truly livable community. Finally, ecologists of an anthropocentric leaning ask difficult questions about how bioregionalists will take account of human interests such as inter-regional equity or minority rights when power is devolved to autonomous communities (Barry 1999b: 86–90). Such criticisms seem to point toward a culturalist bioregionalism. Because "geographical boundaries are not defined solely by nature, but also by human use and perception," argues Don Alexander (1990: 169), it makes most sense to define bioregionalism "in terms of an environmental ethic and a cultural sensibility." Cultural bioregionalists extend the insight of Lewis Mumford (1938), for whom regions reflect geographic, economic, and culture dimensions. In fact, mixed culturalist and nature-centered formulations of bioregionalism are common in the English-language green literature. That is what first led Alexander to conclude that "bioregionalists are somewhat confused" (1990: 164). To overcome the confusion, he offers a "bioregional consciousness" that reflects "the interrelated natural and social realities that impinge on the individual" (ibid.: 172).[8]

Interrelationship does not necessarily transcend dualism. Alexander leaves the two centers intact, for he is not suggesting that what "nature" is is somehow internal to the very meaning of society or individuality. When he says that "nature sets limits, but it does not dictate how we should treat one another" (1990: 169), Alexander barely deviates from Sale's reliance on scientific ecology. He disagrees with Sale not in defining nature but in denying that such science establishes moral *imperatives* for human communities. The "social" side of Alexander's analysis is similarly conventional. He allows people to decide what significance to give scientific information when integrating it into their understandings of a good life. He is not claiming that the very meaning of society or individuality is inseparable from conceptions of "nature." At best, he ends bioregional confusion only by asserting the supremacy of human, communal values. Having rejected biocentric bioregionalism, he opts for a moderately anthropocentric substitute.

Personhood and Nature

At first blush, French personalist regionalism appears to fall clearly on the culturalist side of this controversy. "Regions have to be made; they are not given," says Rougement (1977: 288). Charbonneau too builds federalism "less from natural regions than the alliance of a tradition and its natural setting" (1991b: 157). In some cases, he says, climate and soil conditions determine the boundaries of a region; in others, history, language, politics, or religion predominates. The French theorists show scant interest in connecting regionalism to anything like a philosophy of nature's intrinsic value.

Indeed, Charbonneau (ibid.: 151–152) posts a warning against the very sort of reasoning that supports Sale's bioregionalism:

The temptation of ecology—originally, the science of natural ecosystems—is to limit itself to a naturalism that chooses nature against man, his culture and his freedom. . . . It is a question not of choosing between Nature and Man, but between society that destroys man along with nature and a local society that reconciles them. . . . The regionalist revolt can give the ecological movement its human, social, and political substance. Without this, the ecological movement is danger of getting lost in the ideology and spectacle of naturalism.

In invoking the "laws of nature," Charbonneau would object, Sale effectively "chooses nature against man."

Charbonneau sees the reconciliation of humanity and nature in the *pays* (a word connoting the land, the country as opposed to the city, and village life). The *pays* is a geographical area of common folk traditions, time-honored agricultural and craft practices, culinary preferences matched to local produce, and leisure pursuits fitted to the climate and terrain of the region (Charbonneau 1991b: 156–157). Regions are territories that bear the marks of the persons who have dwelled there.

Yet French personalist regionalism is not identical to culturalist bioregionalism as encountered in the English-speaking literature. Rougement alludes to the difference in a phrase he adds to his observation that "regions have to be made." Regions, he says, "exist as a potential within our needs and desires" (1977: 288).[9] At their most perceptive, personalists regard "nature" in its regional manifestation as something that appears in relation to certain features of the human personality. Their occasional confusions in developing this idea should not keep us from seeing its originality. The concept of the person can yield an ecologism that does not vacillate between the poles of the nature/culture dichotomy.

Long-established personalist ideas take on a greenish hue when theorists confront the consequences of unbounded economic growth. Charbonneau and Rougement point to problems that go far beyond material effects such as resource shortages or the spread of toxins in the environment. Charbonneau (1973: 15) launches his critique "in the name of freedom and democracy." He worries that, faced with shortages of needed goods, scientists and the State will produce substitutes—artificial foods, purified water and air, regimented green spaces—to ensure human survival. Such production will require increasing degrees of organization and social control.

So "science will have liberated man from nature only to hand him over to an even more burdensome society" (ibid.: 160). We are faced with three choices, and only three: "chaos" (wars, economic crises, and penury of resources), "system" (a totally rationalized society), and "an equilibrium, halfway between chaos and system, voluntarily maintained by man, who has become master of his science and tools, as he has become of nature" (ibid.: 15).

For Rougement, confronting the limits of growth raises the question of "finalities." He asks: "Toward what, in fact, is humanity directed in this finite world? Toward progress, as we thought yesterday, or toward the Apocalypse?" (Rougement 1972: 17). He worries that the Club of Rome emphasized material harms at the expense of "psycho-sociological ones" (a generalized fear of the future, unemployment, restrictions on liberty), and that encountering the limits of growth could cause a "penury of the *meaning* of life" (ibid.: 27).

Rougement and Charbonneau see in the ecological crisis further evidence of the need to build theoretically on the critique of individualism that has been a mainstay of personalism since its earliest years. "Individuals" are the social atoms of marginalist economic theory and liberal contractarian theory. They are self-contained and instrumentally rational. Their motives are expected to be materialistic, competitive, and acquisitive. Individuals are essentially ("naturally") free—that is, minimally constrained in the range of activities they may rightly pursue. These characteristics give them great potential for internal consistency. Because pursuing their own interests is their primary motivation, they can imagine arranging those interests rationally so as to allow them to maximize their preference satisfaction. Moreover, the priority they give to satisfying self-originating desires leads them to regard the community as beneficial only to the extent that it helps them fulfill their own interests. To preserve as much scope as possible for individuals to pursue their interests, the community must keep the prescription of communal ends to a minimum. It must leave them free from constraint.

The personalist view of freedom resembles the individualist view only superficially. Individualists and personalists alike refuse to regard human beings merely as sites where anonymous collective forces and social codes play themselves out. According to a personalist account of freedom, however, the psychic coherence of the person consists not in a utility-maximizing strategy of pursuing desires but in the meaningful integration of all one's choices. *Persons* think of moral responsibilities not as restraints on their liberties but as emanations of their very being. They recognize moral behavior as a condition of maintaining community and maintaining community as a condition of genuine personhood (Pierce 1966: 65, 69). They do not hesitate to talk of interpersonal relations as matters of "love," not merely of contract (e.g., Charbonneau 1951: 435). Love unites people through an affirmation of the differences that constitute them as unique persons, rather than through a mutual instrumentalization of capacities that they choose to alienate.

Confronting ecological limits makes the personalists' communitarian critique yet more pertinent—and unavoidable. As long as individualists believed that growth could continue indefinitely, they could argue without inconsistency that people could set their own ends as they pleased and that such freedom would generally not interfere with the welfare of others. The harshest forms of competition between different people's ends could be avoided largely by expanding the supply of goods available to them. The recognition of limits, however, blocks that solution. In a finite world, our ends will conflict more consistently. And the collapse of our "end-less" ways of thinking brings the question of ultimate purposes, unavoidably, back to the fore. Perhaps, as Charbonneau recognizes, humanity can avoid destroying itself by tying individuals down with ever more regulations and then feeding them an array of technologically developed substitutes for the natural goods we have dirtied or depleted. But that does not mean that we have really overcome the limits of growth. Limits reappear as soon as one takes account of a conception of the person as an autonomous being. A regulatory solution operates by shrinking the set of ends available to the individual. If one is obliged to reject such shrinking in the name of human needs for spontaneity, spirituality, rootedness, and beauty, then whatever greater level of natural preservation is consistent with those values *is* the limit of growth. Put slightly differently: Every notion of "natural" limits tacitly contains a view of what human beings need and should be able to tolerate—not only biologically but also spiritually.

The "person" whom Rougement admires opposes every form of homogenizing, collectivist pressure that seeks to shape his or her purposes. The human personality is such that we have a variety of potentials—free activity *and* communal integration, rootedness *and* mobility—that draw us in different directions. Although aware of his spiritual uniqueness, the person also feels solidarity with others (Rougement 1977: 209–211). There is no easy integration of all these ends. For Rougement as for Mounier, personalism *humanizes* individualism (Pierce 1966: 69) by emphasizing the multiplicity of values available to the person beyond those routinely promoted in mass societies.

But the question remains: Does any particular attitude toward the *nonhuman* environment form an essential part of our personhood? When Rougement says that "the regions, *in the last analysis*, are the expression of man's will to self-government [*autogestion*] in his struggle against state-worshipping alienation" (1977: 305), he implies that respect for human autonomy, not "nature," is the ethical ground of personalism. Rougement's environmentalism barely takes him beyond noting that humanity must henceforth use its freedom to take responsibility for the future of the human species (1972: 38).

Charbonneau is philosophically more acute in describing the internal link between personhood and nature. Like Rougement, the Gascon personalist believes that the human personality is inherently contradictory. But Charbonneau explains that "nature" forms one of the fault lines in our divided personality: "Nature is my own sensing and active body, without which the mind would be only an abstract idea." (Charbonneau 1980: 55) We can speak of "nature," first, out of awareness that part of what we essentially are is flesh and blood. We are dependent for our sustenance on a provident world. We are composed of the same biological processes as other living creatures. We are able to feel a oneness with surroundings that arouse in us feelings of beauty and gratitude. Yet the person is also liberty. Liberty describes our ability to create, to negate, and to refuse submission to every externally imposed constraint. Because of liberty, we are also *sur-nature*—beyond nature: dissimilar to other creatures because of our ability to mitigate suffering and to construct a comfortable material world (Charbonneau 1991b: 18).

Charbonneau is at his philosophically most interesting when arguing that these two aspects of our personhood are dialectically interrelated. We need, he argues (1980: 65), the hard, resistant otherness of nature in order to become aware of our own liberty:

It is precisely because nature becomes for me the Other . . . that it exists for the existent that I am. When I no longer reduce myself to nature by making man an element of the universe, or when I no longer reduce nature to man by taking it to be the vulgar matter of his making. When I no longer personify it, as the Ancients did, or, like the Christians, identify it with a Providence that is supposed to satisfy our needs and our reason. Or again, like some naturalists, by endowing it with specifically human rational or moral qualities: such a naturalism too is only a form of anthropocentrism. In order to know nature, it has to be distinguished from oneself: one must love it for its own sake.

The discovery that one is both body and mind—both "nature" and "spirit"—prompts the first awareness of difference (Charbonneau 1951: 434; 1991a: 266–275). Even as we know ourselves to be a unity, we become aware of two parts of our being, each with radically different needs. This is the awareness we develop when learning to respect other people and when seeking to protect our surroundings against the forces of homogenization and destruction.

To accept nature's otherness in this way is to establish a relationship that is properly called a relationship of love. Love is a passionate rapport that seeks union with the other, all the while encountering resistance from it and affirming its difference. It is because liberty and nature intertwine in this erotic way that Charbonneau's ecologism valorizes the *pays* rather than the bioregion. The *pays* is the land that has, over generations, partly yielded to its inhabitants' loving tendance while retaining enough of its own identity to make their freedom meaningful. Thus, Charbonneau's philosophy is not ecocentric. Ecocentrism presupposes an ability to discern "life" independent of our will, whereas for Charbonneau such life appears *because* of our liberty. Nor is he anthropocentric, insofar as anthropocentrists regard human beings as centered in their manner of exercising their will. Utilitarian anthropocentrists, for example, suppose that it is possible to conceive a morally rational integration of human interests, as if they all radiated from a single, unifying judgment. The ambivalent personality experiences no such harmony.

Neither sacralizing nature nor simply instrumentalizing it, Charbonneau's divided psyche lives in perpetual tension with its surroundings: "The relationship of man and space is ambiguous. To live, he needs to dominate and exploit it. . . . But it is also 'his own': habitat is only an extension of the body of the inhabitant, thus, of the spirit." (1991b: 27) Charbonneau's person wants a humanized landscape whose beauty and traditions correspond to the spirit's quest for meaning and whose sustainable productivity corresponds to the body's need for sustenance and security. We should, he suggests, aim to sustain within ourselves an age-old ambivalence between attitudes that sanction exploiting a world we need and preserving a world we love. "The Good," Charbonneau reminds us, "is not in a logic, but in a tension; the Supreme Unity is not in a totalization, but in a plurality that love surmounts" (1951: 435). Ecologism becomes a matter of balancing humanity's instrumentalizing and loving approaches to its surroundings.

Maintaining that balance is difficult. As the examples of Philippe Saint Marc and Jean-Marie Pelt show, some personalists are less successful at it than Charbonneau. Without ever explicitly embracing a centered ecologism, personalists can end up making nature simply a canvas for personal expression or making human freedom simply a variation on natural processes of adaptation. Nonetheless, the writings of Pelt and Saint Marc have the merit of clarifying the arguments for *épanouissement* that pervade French ecologism.

Self-Realization and Nature

Philippe Saint Marc and Jean-Marie Pelt are both academics with Christian-democratic leanings. Pelt, a professor of plant biology at the University of Metz, founded the European Institute of Ecology in 1972 to conduct research on pollution, ecotoxicity, and urban ecology. In the mid 1970s he joined Rougemont's Ecoropa, and in recent years he has been its honorary president (Jacob 1995: 105). Widely cited in the French media and occasionally testifying before parliamentary committees, Pelt is a botanist with a public (but not partisan) vocation. In the late 1990s, for example, Pelt took the lead in opposing the cultivation of genetically modified crops in Europe (Pelt 2000). His books *L'Homme re-naturé* (1977) and *Le Tour du monde d'un écologist* (1990) are among the representative works of French ecologism (Jacob 1995: 157).

Saint Marc has been a professor at the Institut d'Études Politiques and serves as a high functionary in the Cour des Comptes (one of France's prestigious grands corps, charged with assessing the budgets of state agencies). In 1972, he co-founded a new association of some twenty nature-protection societies, the Comité de la Charte pour la nature. The charter, which gathered 300,000 signatures, began by proclaiming that a "right to nature" was the foundation of any civilization and that when using the natural environment we must "transmit to future generations a physical and aesthetic inheritance that they will need for their existence and *épanouissement.*"[10] Consistent with his elite position in the government, however, Saint Marc's politics have otherwise not been notably populist. He is associated with the Centre des Démocrates Sociaux, the tiny successor to a Catholic center party of the 1940s inspired by Mounier's personalism. Since the 1970s he has developed a reputation as a persistent environmental critic of market liberalism.

Saint Marc (1971: 34) asserts that "Man damages Nature because he gains by it—often a great deal—and it is costly to him—often very costly— to preserve it." He argues that nature supplies us with numerous goods beyond those of immediate use value: open space, self-cleansing capacities of air and water, and places for relaxation and contemplation, among many others. Such "immaterial" goods go unpriced or underpriced in economies devoted to maximizing material production. As a consequence, capitalist and communist economies have pillaged their environment (Saint Marc 1978: 387). Since the breakup of the Soviet Union, Saint Marc has targeted "ultraliberalism"—global free trade, deregulated internal markets, ever more individualistic interpersonal relations—as the source not only of environmental devastation but also of a worsening disparity between rich and poor nations, rampant urbanization, rising social aggression, and moral decline (1994: 281, 291ff., 315).

Inasmuch as market liberalism fails to protect the environment adequately, the alternative—as Saint Marc proposes in the title of his most famous book—is "the socialization of nature." "State dirigisme," Saint Marc affirms (1971: 61, cf. 21, 24, 284–298), "is essential; it is up to the State to set development objectives that are compatible with safeguarding the environment and making polluters progressively reduce their damaging emissions." Occasional references to popular participation notwithstanding, Saint Marc shares neither Charbonneau's anarchist predilections nor his sentimental attachment to "nature." For Saint Marc (1971: 43), "the penury of Nature requires that it be subject to planning."

Saint Marc devotes much of his 1971 book to developing a comprehensive "environmental quality index" and a "nature budget" to help state planners allocate nature more rationally by appropriately valuing all of its facets. A more recent work (1994: 425–427) sets forth a program of "ecological Keynesianism": public aid for industries that produce technologies to control pollution or reduce energy consumption, campaigns of preventive health care, investment in green spaces in cities. To strengthen the family, Saint Marc advocates subsidies to allow parents more time at home and the massive development of part-time work. To counter the moral laxity born of individualism, he unhesitatingly prescribes that "a public agency [be] charged with censoring violations of personal dignity: sadism, pornography, criminality—on TV, video-games, and at the movies" (ibid.: 474).

What do such ideas have to do with environmentalism? Saint Marc insists that "the environment" be conceived as the entire life context [*cadre de vie*] in which human activity takes place. He puts the state of rural and urban nature on the same plane as "wild" spaces. He detects similar causes (materialism, productivism, individualism) underlying the degradation of the physical world and dysfunctional social structures.

These concerns give Saint Marc's ecologism a decidedly anti-urban bent. In France, "primitive nature" is too rare for its preservation to be a principal concern of political ecology; better to focus on the benefits of an abundant "rural Nature." In addition to "the biological utility of its forests, its lakes, its rivers, and prairies," the countryside, with its intrinsic beauty and "the harmony between architecture and sites," offers a place for spiritual repose to city dwellers (Saint Marc 1971: 328). Saint Marc would "repersonalize" space, essentially by strengthening rural communities and favoring the dispersion of urban populations into the countryside, where they can once again witness nature's processes directly and dwell in a region's distinctive traditions (1978: 73). Thus "human ecology . . . is not limited to reflection on the physical context but also takes the social and spiritual dimensions of the environment into account" (Saint Marc 1994: 93).

Saint Marc introduces a significant new distinction into personalist views of freedom. For Charbonneau and Rougement, freedom is a metaphysical attribute, an ability to negate existing conditions in the choice of one's ends. Such a view makes it difficult to incorporate matters of social justice into political thought. If every person can resist external determinisms by an act of will, what grounds are there for moral indignation at social patterns of deprivation and oppression? Doesn't every conception of justice suppose that conditions of fear and exploitation wrongly limit the range of human choice and therefore deserve remedy? For Saint Marc, personalist psychology must admit distinctions that support our sense of fairness. It is a fact of life, he observes, that the ability to resist adverse conditions is unequally distributed among human personalities. "In the face of life's trials, people are profoundly different, some are 'fragile,' psychically or physically or often both, others are 'strong.' The inequality of psychic or physical resistance divides the population into two groups that suffer unequally from these trials and who react to them with unequal power." (1994: 59) It is not that the fragile are less worthy beings; on the contrary, the same features of their personality that heighten sensitivity to life's trials often make them more creative and socially generous than people who are less vulnerable. But Saint Marc's distinction does mean that considerations of social justice should influence the distribution of environmental protection. Efforts should be dedicated first to populations in which the most fragile personalities are concentrated (ibid.: 98). In other words, the priorities of environmental protection should be graduated in relation to differences within the psychic constitutions of persons.

Saint Marc names the ethic of his environmentalism "ecological humanism" (1971: 11–12; 1994: 27, 453). Productivist society fails humanistically because it caters one-sidedly to demands for immediate sensual satisfaction. True humanism denotes a commitment to a human well-being that is many-sided. *Épanouissement* requires an environment rich in nonmaterial goods: space for quiet contemplation and leisured appreciation, beauty that elevates the spirit, meaningful work, altruistic engagement in communal projects, and (above all) regular contact with the expansive purity of nature. "Nature is indispensable for the protection of freedom," writes Saint Marc (1971: 185), "because it frees the deep powers of the personality, which are inhibited by the constraints of urban life. . . . [Nature] elicits the creative passion of man. . . ."

Among English-speaking "transpersonal" ecologists, it is common to hear arguments that the human spirit, for its full development, needs exposure to wild nature (Eckersley 1992: 61–63). That cannot be what personalists such as Saint Marc and Rougement have in mind. Both tend to relativize "nature," in its very definition, to the conditions of *human* self-realization. In fact, there is considerable evidence that, in Saint Marc's ecological humanism, "Nature" ends up standing for whatever the multivalent person needs in order to develop physical and psychic well-being. "Nature" is not only pure air and water and diverse forms of life; it is also thoroughly humanized rural landscapes, city parks, tranquil settings of all sorts, convivial social relations, and entertainment purged of offensive spectacles of sex and violence.

The circularity of this type of argument is unsatisfying. It asserts that human self-realization requires a "natural" environment and then defines "natural" as whatever is necessary to human self-realization. This circularity makes Saint Marc stop short of actually defending the particulars of his perfectionist view of the personality. Such reasoning will fail to convince those who do not already share his philosophy of self-realization. Furthermore, relativizing "nature" entirely to the needs of the human personality pays little heed to the unique identity of "natural" phenomena. One could probably have most of the immaterial goods that Saint Marc cares about—pure air and water, green spaces, convivial rural life—in a world stripped of every original forest and populated mainly with domesticated plants and animals. Saint Marc opens himself to Charbonneau's somewhat exaggerated charge (1991: 112) that his conception of "humanized" development tolerates as much artificialization of the environment as that of any real-estate developer.[11]

More attentive to the unique identities of nonhuman life forms is Jean-Marie Pelt, a botanist who has written on the "social life" of plants and on the "languages" through which they communicate with animals. Pelt dwells on the wondrously singular complexities of species interactions. At the same time, in regard to human affairs, he pursues the personalist agenda. Pelt (1977: 140–161, 173, 226, 238, 244; 1990: 467–472) highlights the significance of liberty and responsibility in individual *épanouissement*; prefers regionalist forms of political organization, criticizes unconstrained economic competition, and insists that social justice (particularly in the distribution of employment) is central to political ecology. For him, as for Saint Marc, ecological politics implies "enlarging the field of application of traditional humanist values to nature and to the cosmos" (Pelt 1977: 263; 1990: 457). But where Saint Marc tends to relativize nature to fully realized personhood, Pelt is closer to relativizing personhood to nature—or rather, to systems-theoretic understandings of both nature and humanity.

Pelt depicts systems ecology as a corrective to earlier scientific views of the relation of the person to nature. On the one hand, he argues, earlier scientific views placed at our disposal the means to control natural processes that, to other species, are the unavoidable conditions of life. Other species face disease, deprivation, disaster; processes of evolution regulate their numbers and adaptive responses automatically. We humans respond to nature's challenges by "voluntary regulation," and that makes us "de-natured" creatures. The problem is that, through scientific study, "[man] develops an unfortunate tendency to see himself as autonomous, separated from an environment that he perceives as outside of himself; this development brings about a rupture in the deep links [*solidarités*] that tie man to his milieu" (Pelt 1977: 16). This explains why "the earth has become sick": it has been poisoned by our chemicals, depleted of the biodiversity that constituted its immune system, and exhausted by the "cancerous" growth of megacities (Pelt 1993: 9). On the other hand, science reduces the human subject to a natural object. It attempts to understand complex structures by reducing them to simpler components united in regular, lawful relations. This approach can apply as well to human phenomena as to nonhuman ones. Structuralist anthropologists and sociologists who declare the "death of man" have only carried to its logical conclusion a program advanced earlier through Mendelian genetics and Freudian psychoanalysis. Scientifically observed determinisms lead to denial of the reality of any forms of subjectivity unique to the person. Thus, ironically, the sciences that separated us from the natural world end up reversing the process: "Man begins . . . to see that he shares fundamental determinisms with the animal world and so no longer feels himself radically separated from it." (Pelt 1977: 27)

Neither "reanimalization" nor an assertion of human autonomy from the biophysical world can capture Pelt's hope for a "renatured man." For that, he turns to systems theory. The special intellectual advantage of systems ecology is that it avoids the impasse of a strict opposition between nature and culture. "Between 'nature' and 'culture,'" Pelt asserts (1977: 106), "there are both breaks and continuities: only a dialectical approach takes away the ambiguity of the alternative and empties an endless debate of its passion." In part, Pelt means that systems ecology reveals "solidarities" between humanity and its biophysical environment such that we come to appreciate the need to leave many systemic processes intact. Humanity then no longer sees itself as the external dominator of "nature's" systems; it now sees itself as a part of them. Pelt means something more, however.

Personalists, we have seen, regard the free, responsible individual as the locus of moral concern in the community and as the source of its creative renewal. They worry that powerful, anonymous, large-scale political and economic forces press individuals into a common mold, disfiguring their sense of moral responsibility, depleting the wellspring of inventiveness and energizing commitment that sustains their communities. Personalist ecologists argue that the same depersonalizing trends are deeply implicated in environmental destruction. But unless handled with great philosophical care, their notion of the person's freedom tends to revalidate the traditional view that humanity is external to nature. Pelt's account falls short by this standard.

Pelt (1977: 140–160) argues that systems theory better conceptualizes freedom, by showing how "natural" systems spontaneously generate properties that confound deterministic predictions. The undetermined traits that Mounier and Rougement prized as the essence of the "person"—liberty, individuality, cooperativeness, an ability to transcend limits through creative action—emerge organically, Pelt contends, out of systemically interconnected biophysical processes.

Precursors of freedom appear in the simplest thermodynamic processes. Energy introduced into a physical system at equilibrium can make it pass from a homogeneous to a heterogeneous state—a state in which discernible structures take shape (as when convection currents form in a beaker of heated water). Energy amplifies minute fluctuations in the physical medium, creating macroscopic structures. These structures have some "freedom," in the sense that they maintain their identity against entropic tendencies. But such freedom is minimal; these structures dissipate soon after the energy traversing the system is cut off. Living systems, in contrast, develop homeostatic mechanisms to capture and process energy such that they develop a capacity to maintain their identity across a range of environmental conditions. Even single-cell life forms perpetuate their existence by keeping the concentration of certain molecules in themselves within determined ranges of fluctuation. They are more self-determining than less complex, nonliving systems (Pelt 1996: 58). More complex life forms are also vastly more individuated. Individuation allows each creature to deploy wider arrays of equilibrium-preserving processes—e.g., mobility to search for food, refined senses for detecting the presence of predators. They are "freer."

Human freedom itself can now be seen not as something that distinguishes us metaphysically from every other life form but rather as an extension of systemic properties that appear along a continuum. Life takes form through the complexification of inert matter; consciousness takes form through the complexification of living matter (Pelt 1977: 149). The human brain itself is the product of increasing evolutionary complexity. The neocortex is the seat of specifically human abilities, including "innovation, imagination, creation—in a word, liberty" (ibid.: 253). By developing individuated communities, we extend this process. We collectively create the means to resist—to "free" ourselves from—what other species encounter as environmental determinisms: droughts, floods, diseases, famine.

The problem, according to Pelt (ibid.: 255), is that the same highly evolved cortex that allows us to invent cultural and technological responses to environmental conditions may not be sufficiently integrated with the life-preserving instincts of other parts of our brains so that we can react adaptively to the dangers that our liberty creates. Still, our best hope remains our creative consciousness. If our "automatic behaviors" induce us to ignore or override signs that our technologies are disrupting the natural equilibria on which life depends, then our best chance for survival depends on perceiving those dangers and freely revising our own values as a consequence. In true personalist fashion, Pelt (ibid.: 259) declares that the time has come "for each person to question himself on the meaning of life and on the meaning of his life." We must reexamine our assumptions about our rights of property, about the desirability of accumulating material goods, and about our relationships with our neighbors both in space and in time.

Pelt does not hesitate to apply natural models to society, even as he cautions against applying social models to nature. He criticizes Marx for "politicizing nature"—that is, for interpreting Darwin's theory of competitive relations among species as an ideological reflection of human social relations in contemporary capitalism (ibid.: 22–23). Systems ecologists, in contrast, avoid the partiality of perspective that comes of interpreting nature in human (social-political) terms by studying the roles of both symbiotic and competitive relations in biotic communities. In effect, through systems ecology, Pelt turns Marx on his head, asking rhetorically: "Contrary to Marx, who 'politicized nature,' . . . would it not be appropriate to reinterpret society in the light of nature, thus to 'naturalize politics'?" (ibid.: 234)

Deaf to Charbonneau's warnings, Pelt repeatedly suggests that a renewed human culture take its lessons from the insights of the view of nature offered by systems ecology (ibid.: 157–158, 161, 225–230). He recommends that human communities value individual liberty because, like random "disorders" within ecosystems, it eventually generates new and more successful forms of adaptive behavior. In the same way, humanity should recognize that preserving cultural differences and regional autonomy helps protect the sources of its own renewal. Species maintain not only competitive relations but also cooperative, symbiotic relations. Not to perceive this mix of behaviors in nonhuman communities deprives human communities of models they can use to improve social justice. We must balance economic competition with norms of mutuality.

The idea of naturalizing society in this way remains as problematic as ever. Pelt's science offers something quite different from a nonpolitical model for a program of ecological reform. Thanks to Latour, we can no longer read passages like the following without perceiving how the process of politicizing nature is taking place within them: "In passing from a unicellular to a multicellular being, life takes a decisive step, but at the cost of a heavy sacrifice for each cell integrated into a complex organism; it must henceforth accept the law of the group and renounce the free deployment of all the potentialities contained within its genetic inheritance." (Pelt 1996: 67) "Liberty never consists in being able to do just anything: for a cell as for a human, liberty ends where it impedes the liberty of others. But the sacrifices agreed to [*consentis*] by the cell benefit the entire organism." (ibid.: 74)

In Pelt's nature, cells "renounce" their "liberty," "agree" to "sacrifices," and "accept" the "law of the group." True, they do not engage in a Marxian politics of class warfare. They behave more like tiny Christian democrats! In such passages, Pelt does not take the step that leaves ecologism noncentered. He does not step back from his characterization of "nature" to try to grasp a process that shapes "nature" and subjectivity simultaneously. His portrait of nature ends up being implausibly anthropomorphic. This causes him to skirt an inquiry into whether new forms of political thinking are needed to respond to ecological challenges.

The cautionary lesson learned from an analysis of the writings of Pelt and Saint Marc is that, in green affairs, personalists fall short of their own ambitions when they dichotomize nature and culture. Pelt understands "naturalizing" to mean taking norms (e.g., of symbiosis and of recycling) gathered from observation of the nonhuman world and applying them to human communities. "Nature" remains quite distinct from culture. This opens Pelt to Saint Marc's objection (1978: 229) that "to reduce society to a purely biological vision . . . would be distorting, because it takes too little account of the spiritual dimension of man." Yet Saint Marc's solicitude for spirituality has its own shortcomings. He ends up making "nature" little more than a complement to human self-realization. What gives these thinkers their confidence that human and nonhuman flourishing are ultimately so compatible that no further exploration of these philosophical puzzles is necessary?

Perhaps personalist theology offers one part of the explanation. French personalism draws on religious thinkers such as Jacques Maritain and Emmanuel Mounier. These "realistic" personalists posited two ontic regions open to human understanding: an "ultimate reality [which] is a spiritual, supernatural being" and "a natural order of nonmental being, which although created by God, is not intrinsically spiritual or personal" (Lavely 1967: 108). That distinction cuts through the works of Saint Marc and Pelt. The ecological writings of both men are tinged by Christian faith. Saint Marc's objection to drawing ethical inferences from natural processes depends on his feeling that humanity has a special spiritual vocation, revealed in the Bible.[12] Pelt agrees. In *Le Dieu de l'univers* he argues that there is no necessary incompatibility between Christian faith and the theory of evolution, that scientific study does not reveal God's purposes, and that religious discussion of ultimate ends does not contradict scientific understandings of the processes shaping the biophysical world (Pelt 1995: 172). Theistic personalists write in the tradition of Christian stewardship, which implies that man's "dominion" over the earth is limited by his responsibility to God—that humanity is charged to care for God's Creation, to oversee its general welfare, and not to use it wantonly. Pelt's call to "renew man's ancient alliance with life" and to "garden the earth with love, as in the time of Eden" (1980: 335–336) undoubtedly conforms to such a perspective.

Perhaps it is their faith that exempts these two thinkers from feeling a need to ponder potential philosophical contradictions between the person's autonomy and nature's threatened restrictions on human activity. Their faith holds that all aspects of Creation, as emanations of the Divine Will, *ultimately* fit together harmoniously. It is of little consequence whether we say that nature reflects humanity or that humanity reflects nature. In their common source, the conditions of reconciliation are present. A living world that spontaneously produces qualities of creative liberty and respectful awareness of others generates the potential for human beings to correct their own destructive tendencies (Pelt 1977: 242; Saint Marc 1978: 199–200). God furnishes the materials for making the evolution of the non-human universe and the progress of human society converge. Thus, there is something amiss when Pascal Acot (1988: 241), after reviewing the work of Pelt and Saint Marc, claims that "for ecologists, nature is the supreme value." Clearly, for Pelt and Saint Marc the supreme value is God. In fact, one could say that their ecologism is neither anthropocentric nor ecocentric because it is theocentric: the relationship between humanity and nature is mediated by God, whose purposes unify all of Creation.[13]

The Divided Self: Resisting Power

As the philosophical psychology of Bernard Charbonneau demonstrated, not all personalists glance heavenward when explaining how values transcend personal choice. It seems unlikely, though, that Charbonneau's two-sided dialectic of spirit and body can do justice to the multiplicity of claims that ecologists often make on behalf of "nature." Many greens warn that the greatest danger caused by human activities is not that they will eradicate local cultures but that they will disrupt planetary ecological equilibria. What aspects of personhood account for that more extensive interest? What about claims that certain "w_ld" territories deserve special protection? What reasons are there for contesting the cultivation of genetically modified crops? Establishing the ground of such claims seems to require a more complex psychology than Charbonneau's bipartite personality can furnish. For that, one can turn to the works of Denis Duclos.

Duclos's interest in ecology dates from his university days (the late 1960s) at Nanterre, where he studied under Alain Touraine and Henri Lefebvre, two sociologists who influenced many in the environmental movement. For a time Duclos was a member of the Communist Party. As an enthusiastic young activist in the 1970s, he took stands against nuclear power and dependence on the automobile that eventually set him at odds with a party dedicated to raising material production and protecting jobs.[14] Since the late 1970s, Duclos has been affiliated with the Centre National des Recherches Scientifiques, where his research has focused on the social construction of risk perception.

At first, Duclos was attracted by the hypothesis that risk perception was a function of an individual's specific institutional affiliations and related roles. Gradually, however, he was obliged to widen his perspective. A book titled *Industrialists and Environmental Risks* (Duclos 1991a) demonstrated, unsurprisingly, that factory owners and managers tended to reject ecologism and to defend the social utility of their production. This result complemented Duclos's earlier finding (in his 1981 book *De l'usine on peut voir la ville* [From the factory one can see the city] that workers were hardly more inclined than industrialists to contest socially dangerous practices. In *L'Homme face au risque technique* (1991b: 225 ff.), Duclos argues that making the state responsible for regulating risk is not a sufficient remedy. Where administrative power becomes supreme, additional problematic behaviors set in. Administrators let policies be driven by inappropriately quantified analyses, or they invoke procedures that prevent full or timely airing of complaints. Why are perceptions of risk often handled so evasively? Because, says Duclos (ibid.: 30), a common impulse animates all modern social institutions: "We do everything we can so that the technical machine never stops. . . ." Technologically created dangers can be denied, minimized, quantified beyond all reason, balanced, regulated, redirected—but never declared simply unacceptable.

According to Duclos, social and environmental problems stem from the deployment of various unrestrained forms of *puissance* [power]. Scientific power is unrestrained when its practitioners assume that they can manipulate every part of the world and control any undesired effects with new technologies. The market is unrestrained when all things are commodified and when every part of the world is thrown open to appropriation, development and exchange—even at the price of spreading unemployment, resource depletion, pollution, and safety hazards. Political power is unrestrained when the state acts as if law, surveillance, and punishment were entirely civilizing forces instead of recognizing how often they transmit a desire of control for control's sake and instead of perceiving how administrative procedures end up spreading and masking technological risk rather than reducing it (Duclos 1991b: 251–253).

If our small-scale organizations routinely dodge responsibility for their externalities, and if our large-scale authorities hesitantly regulate them only at the price of attaching distortions of their own, is there any other source of resistance to the "technical machine"? Perhaps. In the course of his interviews with workers, industrialists, scientists, and administrators, Duclos noted responses in which individuals gave voice to reticences or desires that they suppressed when enacting their institutional roles. An automobile manufacturer who attacked ecologists for denigrating the social utility of his firm's products could be found expressing concern about traffic problems in large cities (Duclos 1991a: 166). Engineers and workers who defended their own factory nonetheless spoke of an external world victimized by human aggression generally (Duclos 1991b: 185; 1993b: 115–116). Afterthoughts, contradictions, reluctantly acknowledged fears, and even moral rationalizations reveal that individuals retain alternative frames of reference and critical capacities that escape the control of social norms or institutional loyalties. In those latent critical capacities Duclos detects the presence of a general feature of the personality: "civility." Civility—"the vigilance of persons in their common affairs"—is a vast array of questioning and self-protecting behaviors capable of "moderating the models of action that are imposed on us by the state, the market, and science" (Duclos 1993a: 8).

With his emphasis on a sociability grounded in the individual's feelings, and with his interpretation of civility as a "powerful counter-power" grounded in "individual liberty" (1993a: 306) and even in "love" (1996: 204), Duclos recalls the defining themes of personalism. Like a secular Mounier, he regards concrete individuals, with all their unpredictability and ambivalence, as a source of ethical renewal. Insofar as he criticizes the leveling power of the state as well as the inhumanity of the market, Duclos could make common cause with Rougement and Charbonneau. "The subject's solitude," Duclos admits (ibid.: 32), "increases his fragility in the face of the great ideologies." Yet the subject retains moral significance, because he "alone decides 'where it hurts'": "By experiencing intensely a question of social significance, we do not attempt to resolve it into an ultraschematic pair of alternatives. . . .We recognize the tragic impossibility of the global solution, but we also refuse to withhold comment, to express pain, to demand rights. . . ." (ibid.) "Civility" is not only a form of social competence that acknowledges individual difference; it is also the conceptual complement of a "nature" whose resistance to human appropriation gives meaning to the very idea of respect. Civil ecologism demands an understanding of a self divided along multiple fault lines. Duclos finds such a theory of subjectivity in the psychology of the heterodox Freudian Jacques Lacan.

According to Lacanian theory, the self is not a unified, immediate presence of consciousness to itself; it is not a synthesizing locus of all perception and feeling. Multiple desires, impulses, and longings compete and coexist uneasily within us. The formation of an "I" that can express its consistent identity comes from the outside, first from the perception of the wholeness of one's body in the "mirror stage" of ego development and then more generally from the internalization of how one is seen by others. "In this sense," explains Eric Matthews (1996: 143), "becoming a 'self', becoming fully human rather than merely natural, and so becoming able to gain control over one's desires, is inseparable from becoming socialized."

In Matthews's suggestive interpretation, Lacan implies that "nature" stands for the raw passions of the presocialized ego. These passions never disappear. In fact, their fragmentation remains the fundamental fact of individual existence. The ego's perception of itself as a whole is a fantasy. Still, the construction of a controlling ego changes the passions' signification: they become a threat to the autonomous self. To affirm its unity, the ego turns to the symbolic realm. In the "unity of signification," in linguistically mediated social recognition, it seeks to shore up its identity (Lacan 1977: 126). But in Lacan's structuralist interpretation of language this strategy is doomed to failure because every signifier always refers to another signification—never to some reality beyond language that might fix its meaning. Psychoanalysis helps the ego to deal with desires not by further repressing them but by giving them more direct expression. Such expression constitutes, in a sense, the revalorization of nature.

Duclos's ecologism builds on Lacanian premises, except that for Duclos "nature" is much more than a synonym for the turbulent rearrangement of signifiers inside the psyche.[15] In prose recalling Charbonneau, Duclos (1996: 1) contends that human freedom requires the presence of something real (though perhaps ineffable) outside of us: "Our most intimate liberty depends on the resistance of the real to our enterprises. . . . It allows us not to be alone in the face of the multiple mirror of ourselves: Nature testifies that the loved being has within itself something that is truly other, something not entirely codified by social convention." Liberty presupposes a relationship to that uncodified something. Only the presence of something that it cannot alter keeps the self from drifting in a swirling dreamworld where nothing has meaning because everything can be defined and redefined arbitrarily as a function of pure desire. At the same time, our desires endanger that necessary otherness. Becoming fixated, we interpret everything in our surroundings as a function of one desire or another.

According to Duclos, four "great passions," each originating in a distinct stage of personality development, constitute the fundamental psychic divisions of the human subject. Each mode of subjectivity coincides with a distinct "intuition of nature"—"a propensity to use or destroy it"—that is incompatible with the others (ibid.: 8). The secret to preserving nature is to avoid the seduction of any one of these modes. It is to stimulate free psychic circulation among the passions and their different "natures." Although it is not possible to convey in a brief summary either the elaborateness of Duclos's analysis (which, for each psychic mode, typically posits numerous and sometimes contradictory consequences for "nature") or its plausibility (which depends importantly on one's assessment of Lacanian psychology), any understanding of his version of political ecology requires a survey of the "four great passions."

The first passion consists in our impulse to affirm our identity in the face of otherness. When our surroundings become too uniform, too artificial—too much like ourselves—"nature" gets expressed as that which is "not us." Environmentalist demands for a nature that remains pure and wild are a reaction to social trends that undermine our quest for subjective singularization. Such demands, however, can be pushed to an extreme in which "nature" is asserted as a mute and unalterable whole and in which those accused of altering it become the target of hatred (ibid.: 23–51).

Second, we must constitute ourselves as subjects of objects that we love or reject. In this case "nature" takes the form of a divisible aggregate of qualities and places which the self seeks to master in order to compensate for feelings of missing wholeness. Nature then exists as something "worked over, taken apart, reassembled and then dissociated again" (ibid.: 56–57). This is Cartesian nature: something to master.

The third passion makes people "negotiate the very location of their identities and objects in the collectivity of other subjects," especially by means of elucidating symbols (ibid.: 16). As social relations are made denser, subjecting individuals to norms and laws of all sorts, people cling to their ability to manipulate symbols as evidence of their free subjectivity. The freedom of the socialized subject comes at a price: it makes him "the greatest enemy possible of nature"—that which is outside of all society (ibid.: 74). The symbol-using subject then tries to restore his or her access to the loved object by evoking it theatrically: nature as decoration, as romantic countryside.

Fourth, people must "situate their positions in the system of signifiers that they use to think." In search of a precise content for the signifier, we turn ourselves into objective observers of nature, using precise measurement to discover the rules of its underlying order. Obsessionally, this view of nature can lead to "the attempt to program an 'integrated, global, and interdisciplinary' ecology" in which nothing would escape our desire to reign in every thing that is singular or disorderly (ibid.: 84–87).

In these four passions one can recognize various modes of political ecology: in the first and the second, deep ecology and utilitarian environmentalism; in the third and the fourth, Charbonneau's sentimental regionalism and Rosnay's cybernetic holism. It is not Duclos's intention to declare any of these modes invalid. His purpose is twofold. He seeks, first, to make us realize how these modes grow out of our own subjectivity. Rival theorists

of political ecology—some asserting nature's intrinsic value, others asserting a regionalist environmental humanism—fall all too easily into believing
that what is at stake in their arguments is some all-embracing mode of
thought in which each environmental claim is definitively accorded its
place. Too often we seek to defeat the other's arguments—even if we know,
somewhere in the back of our minds, that our own arguments are incomplete or extensionally implausible. Duclos suggests that psychological
needs, not some ultimate moral Truth, structure such debate. He is saying,
second, that the desire to push any particular mode of subjectivity to its
limit is potentially destructive of human liberty and of "nature." Political
ecology must realize that respect for "nature" is "accessible only by the
detour of a respect for the plurality of human sentiments" (ibid.: 8). Such
respect is the essence of civility.

For Duclos, civility is defined in relation to three other dimensions of
human experience. The first of these is identity: nonvoluntarist social
processes establishing kinship and group membership through which we
distinguish insiders and outsiders. In the second dimension, convention and
law found potentially more inclusive communities through voluntarily
assumed obligations, both political and economic. The third dimension,
science, is also a form of human behavior, not only of cognition. It is an
interrogative attitude toward the world that "rests on a representation of
beings as entities in interaction, organized in various physico-chemical and
biological levels" (Duclos 1993a: 192). Science places the observer in a position exterior to the things observed, ready to disassemble and recombine
them.

These dimensions are all forms of *puissance* that enable human social
interaction. At the same time, each contains an internal logic that, if carried
out single-mindedly, becomes inhuman. To understand ourselves essentially
in terms of ascriptive identities, for example, unleashes fears of the aggres-
sive other, who must then be excluded or destroyed. A completely con-
summated science would treat humanity itself as a will-less object. If such
consummations are avoided, it is because each dimension of *puissance* can
be used to limit the obsessional extrapolations of the others. Thus, science
calls into question assertions about race or ethnicity that might otherwise
turn ascriptive solidarity into paroxysms of expulsion and vengeance.
Recognition of our ability to exercise our will and to form conventional

associations puts brakes on a science that hurtles toward a mastery of all things. Each form of *puissance* is a form of civility in relation to the others.

Thus, civility is not a fourth social logic. It is a noncodifiable "openness" that intuits the limits of every particular form of reason (Duclos 1993a: 221). Civility appears when people, sensing incipient threats from a particular social logic, respond with an absence, a question, a refusal, or an act of resistance. Before investing in any particular ethical system, we manifest civility when we use alternative myths and symbols to play one form of *puissance* off against another. This prevents each *puissance* from achieving a completeness that would spell the end of our ability to adapt to new circumstances and to integrate others into our communities. Thus, civility is an absolutely crucial social competence (ibid.: 28, 238).

Nature and "Civility"

Duclos argues that "nature" is the underpinning of all civility. To restrain oneself from pursuing a *puissance* to its logical conclusion requires becoming sensitive to the presence of something outside of oneself that is not objectifiable, translatable, manipulable, or reducible to a function. "To be civil," says Duclos (1993a: 68), "is to recognize the untamable nature in the other . . . and it is to distinguish it totally from that which we may give ourselves the right to act upon." Duclos is not saying that any specifiable thing or condition is "nature." Quite the contrary.

"Nature" consists in the interstices between the various forms of *puissance*, each of which offers its own partial interpretation of "nature" shaped by its place within a particular symbolic grid. Science, for instance, has typically codified nature in terms of analytically distinct parts that interact mechanically as functions of quantifiable forces; social contractarians depict "nature" as a morally hazardous, precontractual condition in which individuals give free reign to their "animal" passions. As symbolic constructs, none of these interpretations depicts nature as such.

In fact, each symbolic construct is meaningful only because it presupposes a more fundamental notion of nature. If scientists codified everything in terms of mechanical interactions and probabilities, they would deny the reality of the very consciousness that makes it possible for them question the world. Moreover, they would fail to prepare themselves for the all-too-frequent occasions when reality behaves in ways contrary to their theoretical schemas (Duclos 1993a: 85, 90; 1991a: 226). That enigmatic reality is "nature." Contractarians may believe that they transcend the "state of nature" and respect individual liberty by coordinating social relations through consensually enacted law. But a society in which every potential action was morally or juridically regulated would rigidify human relations to the point where passionate expression and gratuitous forgiveness would be frozen out. Without those civilizing actions, liberty could hardly be an attractive social ideal. The source of our spontaneous sociability is "nature" too.

The "nature" to which civility testifies is the ineffable something that "always overflows the determinations of human language" (Duclos 1993a: 68). This sense of nature is the morally necessary symbolic complement of every form of *puissance*. In effect, any *puissance* that manages to approach closure and entirely to control "nature" (as interpreted within its symbolic grid) provokes a more fundamental "nature" to appear.

The promise of ecologism, for Duclos, is that its concern for protecting "nature" can become a means of advancing a civil project in relation to every social logic. "The real object of ecology," asserts Duclos (1993a: 315), "is not purity, but civility. . . . What is essential to ecologism as a cultural phenomenon for the future is ultimately that it supports civility in the form of restraint on acts of *puissance*." What is uniquely important about ecology, according to Duclos, is that it carries the potential for making people realize that each logic of power—and that includes "green" power—stands in need of conscious limitation. It carries this potential because its characteristic preoccupation—the notion of an open, enigmatic, unmanipulable "nature"—is not a *puissance* at all. Civility designates our ability to invoke nature, freely and unpredictably, by moving among the symbolic registers of identity, convention, and science.

This view of ecology leads Duclos to take issue with Serres and Latour for the same reason that the personalist Jean-Marie Pelt rejected Marx's "politicized nature." The fundamental problem is that nature's politicizers are determined to interpret "nature" by means of concepts commonly applied to social relations. For Duclos, on the contrary, nature can perform its civilizing function only if our ideas of it are kept distinct from *our ideas of* society. Ignoring this injunction, those who politicize nature invite disturbing and unintended consequences. Serres's "natural contract" with the entire planet unconsciously partakes of a human passion to surpass our divided condition, to realize a dream of a unified identity (Duclos 1991a: 218–219). That passion is the prelude to a *puissance* that seeks to control everything. For that reason, Duclos fears Serres's goal of "the mastery of mastery." Our fantasms will surely distort that project—an outcome that seems especially likely when we remember that, in the contract tradition, "mastery" implies the use of force to control savagery (Duclos 1996: 149–150; 1993a: 245).

Latour tries to overcome modernity's Cartesian bifurcation of reality. He fuses nature and humanity by investigating the "actions" of "socialized nonhumans" and by demanding that nonhumans be "represented" in new institutions. But he assigns no limits to the creation of hybrids. Having thrown out any concept of nature that might restrict hybridization, Duclos (1993a: 200, 203, 211) argues, Latour unintentionally justifies their *accelerated* production. In an ingenious turn of argument, Duclos (ibid.: 196–198) upends Latour's thesis about modernity and puts Latour squarely in the modernist camp, contending that modernity is constituted not by a division between humanity and nature but rather by the formation of a morally opaque, detached mode of thought allowing observers to manipulate humans and nonhumans alike. To be truly amodern, Duclos argues, one would have acknowledge the limits of social-political constructions of nature as well as scientific ones. One would have to "admit that the real is open, in the sense of being non-communicative, of having an aspect that is 'unbridgeable,' non-constructible, indisposed to being bound into a network, non-assemble-able, and therefore non-dominable" (ibid.: 201). We must conceive "nature" as that which exists beyond any possible symbolic representation.

Even in its demands to respect nature, French ecologism follows a path distinct from that of English-speaking environmental ethicists. Duclos's declaration that "the unalterably wild must exist" (1993a: 309) is not equivalent to English-speaking ecologists' demands to preserve vast territories untrammeled by man. English-speaking ethicists sometimes try to formulate principles to keep human activity within a boundary. Such a limit might be something like Lawrence Johnson's (1991: 276) moral rule of thumb, according to which "we should always stop short of entirely destroying or irreparably degrading any ecosystem, species, or other such holistic entity." Duclos's prescriptions, in every case, amount to a program of multiplying the opportunities for civil interaction—not devising a rule (whether thumbs are involved or not) to curb specific types of human activity. Duclos does not advocate shutting down risky industrial processes; rather, he advocates better integrating workers and scientists into risk-assessment procedures (1991a: 215; 1991b: 251). He proposes not preserving vast expanses of wild territories but rather making cities less exclusively focused on production, on commerce, and on perpetual movements of traffic. Instead, cities could serve as theaters for different forms of passionate involvement: family gardening, playgrounds, places for casual conversation and artistic display (Duclos 1996: 293). Duclos envisages individuals' exercising self-restraint in their consumption decisions because they open themselves to questions about their materialist values (1993a: 310). Civility obtains its anti-homogenizing effects only because it is not bound by rules. Duclos asks for human self-restraint so that power does not control everything. Yet civility picks out no particular thing and no class of things—no species, no territory, no ecosystem—as essential to the maintenance of symbolic openness.

Personalism and Partisanship: A Troubled Connection

Ecologized judgments not only elude definitive formulation as scientific laws or moral principles. For the personalists, they also resist embodiment as an organized political force. In the 1930s, Emmanuel Mounier argued that personalists could join no party en bloc. His focus on the spiritual crisis of twentieth-century man precluded identifying with anything so ideologically prescriptive as a party program. Personalists would be "neither right nor left" (Rauch 1972: 82, 197). Making judgments, personalists agree, is the province of the free individual. Consequently, every institution is suspect. Institutions, as such, seek to regularize the will of their constituent members. Political parties, for example, are not vehicles for self-realization; they exist to advance a collectively binding program. Personalists therefore balk at affiliation with a party—green or not. And if they can be brought to accept the necessity of forming a party, they are likely to insist that it incorporate certain ideological and organizational traits that risk compromising its political efficacy.

Personalist ecologists therefore regularly manifest attitudes that René Dumont would call unpolitical. Saint Marc (1971: 383) argued that "without itself being a party, the defense of Nature must be a pressure group that is sufficiently powerful to arbitrate between parties." For Pelt (1990: 465), ecology has the good fortune of being "simply an ethic, a sensibility, a code of good conduct between human societies and nature"—not an ideology. Ellul (1992: 27) asserted that "ecology has nothing to gain by transforming itself into a political party" and that "the ecological current should develop itself as a counterpower, without playing political games." Similarly, Duclos holds out little hope for an ecology party, noting that a party tends to "get rid of its popular base in order to devote itself to bureaucratic and personality games" (1993a: 10; 1996: 5).

Of all the personalists, Charbonneau is the most emphatic in criticizing partisan ecologism. His reasoning most clearly exposes the tensions between personalism and partisanship. Charbonneau's conviction that ecologism was above all a matter of spiritual transformation led him to prefer the persuasive instruments of journalism to the power-seeking methods of political organizing. He contributed prolifically to environmentalist magazines that were influential in coordinating the activities of the ecology movement in the 1970s. But he criticized Dumont's presidential candidacy, seeing in it a danger that ecology would be coopted in favor of an authoritarian, state-centered approach to politics (Bennahmias and Roche 1992: 38).

In the early 1990s, just when Les Verts had finally organized themselves well enough to achieve some electoral success, Charbonneau protested against the whole idea of electoral ecologism. Competing for power would, he charged, confront an ecology party with a terrible dilemma. If it won power, the party would inevitably betray its ecological principles. To get enough votes to win, it would have to avoid attacking the productivist system frontally. It would have to be satisfied with creating a natural park here, a more stringent pollution control law there. No electorally successful party could do what is really necessary: challenge society's addiction to economic growth. If it remained true to its anti-productivist principles, the party would be impaled on the other horn of the dilemma. It would find itself in the opposition, all the less able to challenge productivism for being alienated from the other opposition parties. The questions that ecology raises, Charbonneau concludes, are too socially unsettling to have an electoral payoff. "Therefore," he argues (1992: 39), "they have to be asked from the silent retreat of an institution that has escaped from power struggles, from a place where debate is not reduced to competition between media stars for the conquest of the public." In today's world, it is hard to imagine such a retreat outside of a monastery or convent. The media intrude everywhere. Markets and state power constantly impinge on our choices. Launching uncompromising criticism from some "silent retreat" may be satisfying to the soul, but it does not reverse environmentally destructive trends that those forces accelerate. That is why Dumont tirelessly insisted that ecologists seek enough power to enable effective opposition to such forces. Yes, a party may betray its principles as it "plays the political game." But the success of *political* ecology surely has to be measured in the effect it has in securing sustainable social-productive practices in a more just *world*. As ecosocialists have long taught, any hope of achieving those objectives requires taking the risks of organizing and exercising power.

Nonetheless, personalist values of nonconformity and regional identification have had a profound impact on the French ecology movement. For better or for worse, they helped delay the creation of an ecology party until 1984. By then, green parties already existed in at least nine other European countries. Once Les Verts was created, personalist ideas of the importance of "autonomy" helped stoke tensions over whether the party really aimed at power or whether it was more a channel for protest and "convivial" activities (Sainteny 1991: 29–31). Personalist values supported the formulation of party statutes that dispersed power: no individual would be designated party leader; regional organizations, not the national executive committee, would choose candidates; representatives would not receive voting instructions (Boy, le Seigneur, and Roche 1995: 94–99). Perhaps most fateful, personalism underlies Les Verts' claim to be "neither right not left"—a slogan associated most strongly with the rise of Antoine Waechter.

Waechter's insistence on marking ecologism off from the socialist left is, in fact, an additional characteristic of personalist ecologism. Most personalists are ill at ease with egalitarian collectivism, with materialist interpretations of history, and with conceptions of political "struggle."

Ecosocialists, on the other hand, are dissatisfied with a vision that downplays the significance of "material" forces in forming the individual will and that sometimes weakens political action on behalf of issues of social and international justice. These misgivings have not, however, stopped ecosocialists from absorbing some personalist ideas. If French ecosocialists are more "libertarian" than the traditional left, if they speak more of decentralization and *épanouissement* than of central planning and class identity, it is because they have forged their ideas in a dialogical context where personalists—not ecocentrists or anthropocentric environmental progressives—have been among their consistent interlocutors.

6

Socializing Nature

It is obvious, Serge Moscovici once asserted (1978: 138), that the ecological movement has to be situated "in the socialist stream" because of its desire for "a more communicative, more just, and less hierarchical society." At least it was obvious in France, where most major theorists of ecologism have had a leftist past. Even advocates of seemingly distinct forms of ecologism—those indebted to the systems approach and to personalism—have adopted a language of "socializing" nature (Rosnay 1975; Saint Marc 1971). The first thing to strike an English-speaking reader about ecosocialism in France is not the novelty of its themes but the pervasiveness of its influence.

As a specific current within the stream of green thought, French ecosocialism was launched in the 1970s, when André Gorz and René Dumont played major roles not only in politicizing ecology but also in drawing attention to links among resource depletion, alienating work conditions, and unjust treatment of Third World countries. Later ecosocialists followed suit. In the 1990s, Alain Lipietz (1993a: 37) argued that green thought "refounds" Red theory by siding with the exploited, workers, and the Third World. Likewise, Jean-Paul Deléage, the editor of the journal *Écologie et Politique*, contends that "without the socialist idea, there is only an ecology that is nostalgic for a bygone past" (1997: 138).

No one should take the durability of the ecosocialist label as a sign of theoretical homogeneity or intellectual stagnation. In fact, French ecosocialism has undergone a marked evolution. In the 1970s it was the apprehension of seemingly objective limits of growth that ecologized the political sensibilities of socialists such as Gorz and Dumont. They responded with unabashedly utopian theorizing. In the process they created unresolved tensions between centralizing and decentralizing approaches to ecological reform. Efforts to decrease those tensions by conceiving of politics as consensus building were stymied by a long-standing Marxist prejudice against contractualism. Marx believed that contract theory was to the bourgeois epoch what natural-law theory was to feudalism: an intellectual mystification that conceals how the state functions as an instrument of the interests of the ruling class. In the 1980s, apprehensions about the class-bound, homogenizing effects of any notion of consensus caused Pierre Juquin and Félix Guattari to try to propose a relativistic ecosocialism.

Since the 1980s, some French Marxists have begun to take a more sympathetic view of contractualism. Such sympathy has strengthened a current that I call "contractual ecosocialism." In contract theory, one regards the terms of cooperation under which a community lives as results of collective deliberations among equals. Ecosocialists such as Lipietz and Deléage widen the range of value pluralism in contractualism by throwing environmental concerns into social negotiations. In so doing, they depart from Marx's most utopian belief: that scarcity—and with it, the social "contradictions" that give rise to politics—might be abolished. At the same time, they modify even their original perception of earth's finitude. Ecological limits are no longer encountered as absolute constraints imposed by nature. Environmental dangers come into focus only through a critique of the value-leveling power of economic rationality.

Of all the varieties of ecologism, ecosocialism has been the most successful in generating a truly international dialogue. English-language studies of ecologism acknowledge the influence of André Gorz far beyond France (Dobson 1995; Eckersley 1992; Atkinson 1991; Pepper 1993). The main ecosocialist journal in the United States, *Capitalism, Nature, Socialism*, has offered Lipietz and Deléage a channel of expression in the English-speaking world. The success of ecosocialists in sustaining a cross-national conversation means that French ecosocialism has more in common with its ideological equivalent in the English-speaking world than other varieties of French ecologism have.

Nonetheless, cross-cultural theoretical comparisons pay off even in this case. English-speaking ecosocialism competes with other forms of ecologism that tenaciously mark out distinct identities: deep ecology and allied forms of nonanthropocentrism, social ecology, ecofeminism.[1] Those currents of green theory barely trickle through France, and this difference significantly influences ecosocialist expression. Confrontations with non-anthropocentrists have driven some English-speaking ecosocialists to accept their rivals' terms of debate. The French contrast, I shall argue, shows why this acceptance is unnecessary. The feebleness of nonanthropocentric ecologism in France has left French ecosocialists freer than their English-speaking counterparts to develop lessons from Marx in ways that dissolve the significance of any alleged center of environmental concern.

From Marx to Ecosocialism

Ecosocialists usually pay homage to Marx while finding historical materialism theoretically wanting in relation to ecology. That is why the green thinker Alain Lipietz (2000: 74) declares that "virtually every idea of Marxist thought must be thoroughly reexamined in order really to be of use." On the one hand, some greens find Marxism attractive because it reveals the inner mechanisms of productivist society. Ecosocialists regard the ecological crisis essentially as a worsening conflict between continuously expanding industrial societies and the earth's limited biological carrying capacity—a conflict exacerbated by unequal relations between developed and developing countries (Ryle 1988: 1–2). Better than any other theoretical framework, Marxism explains how a socioeconomic system that valorizes endless buying and selling (rather than the "satisfaction of definite wants") provokes an unprecedented acceleration in the transformation of resources into waste (Deléage 1994c: 44). Marx and Engels even displayed some green precociousness in their sensitivity to what would today be called environmental effects of production for profit—e.g., pollution and soil depletion (O'Connor 1998: 122–123). In a passage much favored by ecosocialists, Marx (1867: 25) recognized that nature has some value independent of its instrumental value for man, and that labor is the father of use value while "the mother is the earth." In addition to perceiving common ground on environmental matters, Reds turn green when they discover kindred social concerns in the ecology movement. Lipietz (1993a: 11) evokes these concerns with his claim that "today, being genuinely faithful to the struggles of the oppressed requires developing the branch of the green paradigm that favors social welfare and global solidarity." Ecosocialists enlarge upon the notion that production should be geared directly to human need so that ecological considerations become integral parts of production decisions (Dobson 1995: 171). Greens generally agree with Marx's socialist vision too in their championing an egalitarianism that opposes differentials of class, race, and gender and in struggling to make work and social life more varied and meaningful. Marx's understanding that capitalism and its successor modes of production would eventually overcome parochialism, fostering a sense of solidarity in the human community as a whole, resonates with greens who emphasize the planetary dimensions of our ecological predicament. On the other hand, ecosocialists commonly acknowledge that Marx frames his emancipatory ideal in a way that stunts its ecological potential. Marx usually assumes that capitalism is constrained only by its *internal* contradictions; he does not acknowledge there are *external*, "natural" constraints on production (Deléage 1994c: 47). Aside from exceptional passages, Marx tends to assume that, until labor transforms them for human use, the means of production supplied by nature are valueless. And his historical eschatology turns the succession of modes of production into a tale of human emancipation *from* the constraints of nature.

Socialism requires an "eco-" prefix because Marx's hope to meet human needs through ever-expanding, socially rationalized production did not take account of the earth's finitude: its finite supply of resources, its limited ability to reabsorb waste and restore biologically depleted ecosystems. This is what led James O'Connor (1998: 165) to hold that Marxism, to become ecological, must recognize the "second contradiction of capitalism," in which "the combined power of capitalist relations and productive forces self-destruct by impairing or destroying rather than reproducing their own conditions." French green theorists are largely in agreement with this. Ecosocialists recognize that "the age of a finite earth has begun" (Lipietz 1989: 61).

To an English-speaking ecocentrist, ecosocialists look unambiguously anthropocentric. Ecosocialists direct their ethical concern not to nature's "intrinsic value" but to nature's role in sustaining human well-being (Eckersley 1992: 127–132). A determination to surmount social inequalities, to improve working conditions, and to overcome world poverty is all well and good. But concern for the well-being of nonhuman entities per se seems to be beyond the compass of socialist ethics. By now there should be nothing astonishing about that complaint. But there is reason for consternation when an ecosocialist takes the point. Unrepentant, David Pepper (1993: 232) adopts that label: "Ecosocialism is anthropocentric (though not in the capitalist-technocentric sense) and humanist. It rejects the bioethic and nature mystification. . . ." Other English-speaking ecosocialists try to edge closer to nonanthropocentrism, but in the end they seem unable to avoid the dichotomizing language that the debate thrusts upon them. Arguing that "care for the natural world is constitutive of a flourishing human life," the left Aristotelian John O'Neill (1993: 24) denies only that his view is "narrowly anthropocentric." Ted Benton (1995), who pushes ecosocialists to recognize obligations toward animals, believes that "anthropocentrism is necessary" to correct misanthropic tendencies in deep ecology. One ecosocialist, Luke Martell (1994: 11), "rejects anthropocentrism" in favor of a socialist ethics that respects the intrinsic value of animals. Martell's move only underscores the point that, in order to engage with the arguments of nonanthropocentrists, theorists feel bound to launch their ideas from one center or another.

That so many *ecosocialists* choose to argue in this way is puzzling because they have every reason to take issue with a debate cast in terms of subjective or objective centers. Here I point out only what these thinkers already describe perfectly well in their works. Pepper (1993: 107) declares that "Marxists offer a dialectical view of the society-nature relationship [in which] humans and nature are part of each other. . . . Indeed, they *are* each other: what humans do is natural, while nature is socially produced." Benton (1993: 72) builds on the early Marx's notion of humanity's "metabolism" with nature; he places views of the moral status of animals in the context of the evolution of labor in the direction of increasing urbanization and industrial forms of organization.

The question that must be asked here is simply this: Why characterize these processual interpretations in terms of *centering* at all? If labor transforms our understandings of "nature," then "respecting nature" has no meaning independent of "nature's" significance at a particular stage of social development. Moscovici commented 20 years earlier on the passages in Marx that attract Benton's attention. Moscovici (1974: 127) argued that Marx's early writings revealed a thinker for whom "man himself and his work form an integral part of nature and a humanized nature is a nature transformed by human work, a nature associated with the human body and mind." No centrism can be salvaged from the lesson that Moscovici (ibid.: 128) drew: "Outside of this confrontation [between human industry and nature,] nature is an undifferentiated chaos. . . . Strictly speaking, it is nothingness." The qualities on which nonanthropocentrists center their theories of "nature's" intrinsic value (e.g., homeostatic interrelations in ecosystems, sentience in animals) are results of a "confrontation," not simply properties of the "natural" object. Thus, treating them as if they had an identifiable core is a matter of ideological abstraction.

Correspondingly, if human consciousness and values—even our very senses—develop in relation to an environment whose qualities take shape through labor, the reflecting subject has to be understood, too, as a moment in a history of "confrontations." No center there, either. The legacies of colonialism and of class, race, and gender discrimination further proliferate perspectives on nature. For reasons such as these, it makes sense to get away from the language of theoretical centering, as a few English-speaking ecologists who are sympathetic to ecosocialism actually suggest (Dryzek 1995: 18; Guha and Martinez-Alier 1997: 95).

In France such corrective action seems superfluous. Almost since the moment that ecology became politicized there, ecosocialists have taken for granted the theoretical move that keeps ecologism off center. They have focused on understanding the dynamics of historical, processual relations between nature and humanity, rather than on reasoning from a particular definition of either term.

Utopian Ecosocialism

Early on, the men most responsible for politicizing ecology in France, André Gorz and René Dumont, gave ecosocialism two distinctive features. Dumont's ecologism is uncompromisingly Third Worldist. His outrage at international inequalities shocks greens out of forming political priorities as a function of ecological problems encountered mainly in highly industrialized societies. Gorz's writings, on the other hand, epitomize attempts to link outrage at environmental destruction with a commitment to democratic self-management and to limiting the spread of economic rationality.

Gorz was already a well-known figure of the independent French left before he became involved with political ecology. Austrian by birth, he settled in France at the end of World War II. He was close to Jean-Paul Sartre in the postwar years, and he became a prominent voice in left-wing journalism. *Le Nouvel Observateur*, *Le Sauvage*, and other magazines carried the essays that became reference works for French ecologists in the 1970s. These essays built on earlier books in which Gorz assessed the prospects of a resurgent radical politics that would give workers control over both factories and society at large. In *Stratégie ouvrière et néocapitalisme* (1964), Gorz argued that the change in the composition of the working class caused by the shift from manufacturing to white collar jobs did not put an end to the alienation of labor. The "new working class" sought not only better wages but also more room for creativity and responsibility on the job. Such aspirations had received a boost from a crucial development in the evolution of capitalism. The spread of automation, Gorz believed, would enable labor to impose more of its ends on the work process than had been possible during capitalism's earlier manufacturing stage. Automated capitalism creates the conditions for *autogestion*: shifting power over working conditions, technologies, and production goals from managers to workers within capitalist enterprises.

In his subsequent work, Gorz can be read as attempting to preserve as much as possible of this strategy even as he acknowledges additional constraints on it. By the early 1970s, he no longer believed in the emancipatory potential of the new class structures. In *Critique de la division du travail* he found that technical-scientific workers simply "produce the means of exploitation and oppression of manual laborers and must therefore appear to be agents of capital" (1973: 263). His earlier analysis had failed to account for the politics of technological choice. Capitalists, he now charged, chose technologies not only to maximize productivity but also to ensure that workers could not gain control over production. Overcoming capitalism's tendency to disempower labor would therefore require "post-industrial" technologies—technologies operable without hierarchical supervision, suitable for localized production, and easily adaptable to the users' own ends.

Gorz's theoretical revisions implied that capitalism, although it reinforced social inequality and frustrated workers' aspirations for autonomy, failed to generate contradictions powerful enough by themselves to motivate system-transforming social movements. His discovery of the earth's ecological finitude, however, suggested new possibilities for overcoming capitalism. Gorz argued in the 1970s that the profitability of capitalist production had always depended on consuming unpriced resources such as clean air and water. As pollution grows worse, however, firms are forced to pay for those resources—for example, by installing pollution-control devices. This increases the "organic composition" of capital. Likewise, the increasing scarcity of raw materials requires additional investment to search out new natural resources or to process old ones more thoroughly. Yet "the economic effort to overcome *relative* scarcity causes—once certain thresholds are passed—*absolute and insurmountable* scarcity" (Gorz 1977: 23). Prices of goods then increase more quickly than consumers' purchasing power. As consumers purchase less, production stagnates or declines, thus deepening economic recessions (Gorz 1975: 12). Ecological limits therefore feed into the recurrent crises of capitalism's relations of production. Not that worsening diseconomies of the capitalist exploitation of resources will necessarily cause capitalism to collapse. Gorz believed capitalism to be quite capable of a "technofascist" adaptation, in which ecological engineers would centrally plan production to avoid crossing ecological thresholds. But socialists could use ecology as a "lever" to promote social transformation to a self-managed society.

A passing remark by Gorz became central to all his later work: To achieve a social system in which human autonomy is realized, he said, "it is necessary to break with economic rationality" (1975: 25). Since the 1970s, Gorz has come to see capitalism's longevity depending less on an authoritarian technocracy and more on the self-policing behavior of a consumerist population. The extension of consumption is part of the logic of economic rationality: "The quest for maximal economic returns . . . consists in selling the largest number of goods, produced with the greatest efficiency, for the highest profit possible. This requires maximizing consumption and needs." (Gorz 1991: 91)

Contemporary capitalism, according to Gorz, allays its problems of over-accumulation and unemployment by extending this logic into new areas of life. Functions once left to individuals and families—care for the elderly, early schooling, leisure activities—become "colonized" by profit-seeking organizations and thus monetized, routinized, and depersonalized. The "human resource" ideology enables people to adapt to an ever-increasing pace of economic and technical change. It justifies integrating a new managerial elite into profit-making institutions; it encourages consumption of a wider variety of services. The spread of economic rationality stifles the growth of "ecological rationality," which "consists in satisfying material needs as well as possible, with the smallest possible number of goods that are highly useful and durable—thus with a minimum of labor, capital, and natural resources" (Gorz 1991: 91).

To break with economic rationality, Gorz proposes conceiving society as composed of distinct spheres. In only one sphere—that of heteronomy—would criteria of economic efficiency continue to hold sway. Gorz concedes that if the material needs of a large human population are to be met, and if advanced technologies are to be allowed to reduce the time allotted to tedious forms of labor, then some forms of "functional integration" of individuals—including externally prescribed norms and hierarchy—have to be used. The heteronomous sphere "assures the programmed and planned production of everything necessary to individual and social life, with the maximum efficiency and the least expenditure of effort and resources" (Gorz 1980: 97). Planners would see to it that such work was kept to a minimum, that its burdens fell equitably across the population, and, presumably, that production processes respected ecological norms

Scope for more autonomous activity would be found in the two other spheres. In the sphere of private activity, people could develop their talents and interests, for example in writing or art. In an intermediate sphere, individuals would gather in small, self-managed groups to engage in artisanal production or cultural activity. Such a society would embody the principle of measuring wealth "not in terms . . . of exchange values but the possibility of the self-determination of happiness" (Gorz 1980: 124). Gorz's hope rests with the possibility of expanding the sphere of autonomous activity— especially by reducing work time—thus shrinking the space in which economic rationality reigns supreme. Where "Anglo-Saxon environmentalism" simply puts new constraints on whatever products and practices capitalism happens to generate, Sintomer (1991–92: 107–111) argues, Gorz's ecologism aims to reduce the scope of exchange itself so as to open individuals to new social and cultural activities.

One can read Gorz as an anthropocentrist, and that is what interpreters who insert him into the field of English-speaking ecologism usually do (Eckersley 1992: 129; Little 1996: 53). Gorz's worry that "absolute scarcity" will eventually cause economic crisis seems clearly human centered. Still, unlike some English-speaking ecosocialists mentioned earlier, Gorz himself never confesses to anthropocentrism and never bothers to refute ecocentric objections. Thus, we should be cautious about imposing the terms of this debate on his work. In fact there are good reasons to balk at this imposition. After the 1970s, Gorz backed off of his early formulations about the earth's finitude confronting humanity with "absolute scarcity" (Little 1996: 51). His more recent thesis regarding the spread of economic rationality suggests that ecological limits appear in a historical process that shapes both our understanding of nature and our self-understandings. Gorz insists first that the "nature" he wishes to protect is not the "objective" nature of scientific ecology. The functioning of *that* nature, he asserts, could be ensured only by recourse to the technical control of an "expertocracy" (Gorz 1993: 58). Gorz also now essentially admits that, in its own way, capitalism copes with scarcity. It puts (and raises) prices on ever more goods (now including water and air); it encourages technological innovations that reduce consumption of scarce goods; it can even favor regulatory schemes that preserve its popular standing by forcing it to clean up clear health hazards. "Nature" does not impose absolute limits on human activity, but all "overcoming" of ecological limits is accomplished at the cost of allowing economic rationality to destroy the valuational resources of the "lifeworld."

In the phenomenological tradition in which Gorz's thought is located, the lifeworld is the existential correlate of "body-subjects." Unlike a disembodied mind that calculates and analyzes data presented to it, the body-subject is a presence to the world that perceives and elaborates meaning.[2] It is predisposed to form affective attachments and to create multiple settings for the expression of qualitatively distinct types of creative activity and social cooperation. Body-subjects are beings with an orientation to find and build on conditions in the external world that are essential to the development of their autonomy (Bowring 1995: 78–79). The idea of the body-subject helps to build a bridge between consciousness and nature by taking account of the ways in which there is a proto-consciousness—an orientation to discern intelligible patterns—in matter itself.

In the lifeworld, "nature" refers to all the ways in which human beings experience and value their surroundings before their perceptions are rationalized by scientific, economic, or even philosophical methods. Lifeworld perceptions may well include brute dismay at a blanket of brown air over a city—even if experts offer reassurances that pollution levels are tolerable. As inhabitants of a lifeworld, we may be awestruck at an alpine landscape—even if economists argue that the area would be economically more valuable as a ski resort. In everyday life, we lavish affection on animals—even if such sympathy seems ethically discontinuous with our carnivorous diets. Out of the lifeworld comes a multitude of orientations that, no matter how confused from a philosophical viewpoint, often have a validity that intellectual systems ignore. These orientations play a crucial role in stimulating individual creativity and social criticism. As economic rationality colonizes the lifeworld, it destroys such sources of cultural renewal—including orientations that favor environmental concern. Conversely, to appreciate the lifeworld and to defend its valuational resources is to make ecological limits stand out in relief. That is, if capitalism's rationalizing responses to environmental dangers are blocked by values drawn from the lifeworld, we must devise new social strategies to adapt to "nature." So this nature exists neither as an enveloping system from which we draw value nor as a separate entity that humanity learns to appreciate in more diverse ways. Its existence takes shape acentrically, in relation to a shifting field of values.

None of these refinements in interpreting Gorz's theory immunize him against political criticism. Some scholars charge that Gorz's attempt to mix decentralized communities and central planning is fraught with inconsistency. As long as complex technologies are used, and as long as communities are woven into webs of exchange, autonomy will be compromised (Frankel 1987: 59–61). However, it is too rarely pointed out that Gorz's mixing of themes of autonomy and *ecology* is even more problematic. The answer to the question "In what sphere is ecological responsibility handled?" must be "It is almost entirely a heteronomous function."

Gorz asks us to imagine a society in which goods to meet mankind's fundamental material needs are produced in highly automated enterprises, coordinated by state planners. Thus, the sorts of economic activity that have led to environmental destruction under capitalism—mining, large-scale agriculture, chemical manufacturing, and so forth—will continue in some form. If they are no longer to be environmentally damaging, it must be the case that planners, not "autonomous" workers, take responsibility for adjusting production decisions as the earth's ecological health requires. Moreover, Gorz cannot simply assume that autonomous enterprises will be ecologically responsible. He has argued that workers in self-managed enterprises would spontaneously resist the tendency of capitalism to increase production and to stimulate needs (1993: 55–67). Even if that doubtful premise is true,[3] it leaves untouched the ecological problems that may arise from such matters as choices of production technology or the synergistic effects of seemingly harmless effluents from different enterprises. Autonomous producers will need heteronomous regulation to control such matters as siting, choice of chemicals, and waste-disposal techniques. That function too must devolve on the planners.

Finally, although Gorz often expresses concern for "the Third World," it is not at all clear how reorganizing work patterns in "the North" will bring food and environmental recovery to "the South." In that regard too, the heteronomous sphere of affluent societies is the only recourse. Since it oversees the mass production of necessary goods, it must take responsibility for raising the standard of living in the Third World until the Third World can do so itself. Citizens of the affluent North might well face considerably more "heteronomously" mandated work than Gorz believes. The thought of France's other founding ecosocialist, René Dumont, complements Gorz's by facing up to some of these consequences directly.

Earlier we saw that Dumont played a significant role in politicizing ecology. His book *Seule une écologie socialiste* leaves no doubt about the ideological synthesis that underlies his politics. Dumont insists that "today what is bringing about the destruction of ecosystems is the pursuit of profit posed as a law of development" (Prendiville 1993: 34). The inegalitarian effects of "the inherent capitalist priority of profit . . . are even more true at the international level," says the English ecosocialist Raymond Williams (1995: 54). Dumont (1978: 187) turns that rather detached observation into a blazingly passionate Third Worldism, asserting that Third Worldists "battle to end the desperate inequalities and pillage of the Third World"; they decry the plight of inhabitants of countries with high rates of poverty and illiteracy—countries whose low rates of industrialization make them dependent on exporting raw materials to earn foreign exchange. Dumont, who has spent his entire professional life documenting the effects of poverty and environmental devastation in Africa, Asia, and Latin America, sees his contribution to ecologism consisting mainly in "completing the theses of the Club of Rome . . . by insisting much more on the *tragedy* of the Third World" (1986: 12).

In Dumont's view, most of the responsibility for the "tragedy" of the Third World falls on the industrialized countries of North America and Western Europe. First, this is because "problems like the exhaustion of raw materials and pollution are not the doings of the poorest people, but come essentially from the richest countries" (1973: 46). Prodigious consumption of petroleum, emission of ozone-destroying refrigerants, demand for tropical woods, and similar problems emanate from the countries with the world's highest standard of living. Second, developed countries must make amends for a long history of unjust relations with the Third World. Colonial powers deliberately overturned local agricultural practices in order to get products such as coffee and sugar. Many of those economic structures remain today, shoring up local elites and ensuring the "independent" countries' continuing trade dependency on their former colonizers. Third, even in the post-colonial era, First and Second World economies retain unfair trade advantages over the Third World: "In the framework of what we today call unequal exchange, we buy . . . raw materials and agricultural products from underdeveloped countries at low prices, and we sell them back finished products at high prices." (Dumont 1977: 239) Unequal exchange brings together the conditions for the "pillage of the Third World." Unable to earn enough foreign exchange to finance their development, these countries are trapped in poverty. Thus, if their inhabitants are pushed to adopt ecologically destructive practices (e.g., deforestation or the cultivation of soil-destroying crops) just to survive, the responsibility is not primarily theirs. In many ways, developed countries are implicated in sustaining "an intolerable world" (Dumont 1988).

The distinguishing feature of Dumont's ecosocialism is its requirement that we assess the state of the planet from the perspective of "the exploited peasantries" of the world's poorest countries (Dumont 1988: 273). When politicizing the French environmental movement, it will be recalled, Dumont refused to allow the worries of local preservationists to set the agenda. By continually linking problems of environmental degradation with capitalist strategies of economic development, unfair trade practices, and demands for the equitable distribution of burdens in overcoming global ecological problems, Dumont anticipates what Guha and Martinez-Alier (1997: 18) call an "environmentalism of the poor." In such a perspective, those authors explain (ibid.: 95), debates between centered ecologists miss the point: "The two fundamental problems facing the globe are (1) over-consumption by the industrialized world and by urban elites in the Third World and (2) growing militarization. . . . Neither of these problems has any tangible connection to the anthropocentric/biocentric distinction. Indeed, the agents of those processes would barely comprehend this philo-sophical dichotomy." Were Dumont ever to enter so abstract a controversy (something that his penchant for concrete activism has never allowed him to do), his opinion would likely be no different.

In effect, ecocentrists who criticize the "anthropocentric" instrumental-ization of nature dilute responsibility for environmental devastation. That adjective puts an Indonesian peasant drawing a meager existence from cul-tivating rice on the same moral plane as a Western consumer of rare woods taken from rainforests. An ecosocialist rejects such ethical reductionism. But simply calling Dumont an anthropocentrist seems inadequate too. That label grossly understates the difficulty of taking seriously the range of environ-mental concerns that are already present in communities around the world. Dumont's polemical habit of juxtaposing consumption in wealthy countries with the conditions of daily life in the poorest ones is designed to jar Western readers' comfortable suppositions about what is "natural," normal, or toler-able. In that sense, at least, he nudges us and throws us off moral balance.

What should be our response to exploitation when we finally perceive it for what it is? At times Dumont's socialist inclinations, combined with a perception of the severity of humanity's environmental problems, have led him to advocate precisely the sort of planning-based policies that Gorz (at least in theory) tried to minimize. In the early 1970s, Dumont imagined that the transition to ecosocialism could include the state taking control of the "commanding heights of the economy." He even foresaw the need for a "planetary organization provided with real powers" that would decide a "centralized allocation of scarce resources" (Dumont 1973: 95,106). More recently, however, he has become increasingly pragmatic and has placed his faith more in democratic negotiations than in centrally commanded regulation to achieve socialist objectives. Consider some of his most recent proposals: Foreign aid to Third World countries should favor "human development." Investment in education, vocational training, and preventive medicine should take priority over investment in industry and infrastructure. Reforestation should be encouraged and debt should be reduced through debt-nature swaps. Land reform is essential. Creditors should ensure that economic power is channeled to women as well as men. Where interests of workers in the North and the South are in competition, as in the textile industry, governments in the North should tax imports (thus protecting their workers) and should then use the tariffs to promote social and environmental protection in countries of the South (Dumont and Paquet 1994: 122–136).

Dumont's Third Worldist program, like Gorz's autogestionnaire project, remains radical. Carrying through such policies in ways that would effectively rebalance North-South relations would require from the North long-term commitments and financial sacrifices that no existing government has come close to endorsing. What has diminished among French ecosocialists is the utopian impulse. That impulse takes theory beyond positing certain ethical ideals. It sets about imagining their complete realization in a thoroughly revolutionized society. Gorz concluded *Écologie et liberté* with a scenario in which France has been magically transformed. There are workers' cooperatives and liberated schools. Bicycles and trams are provided for transportation. Industries are ordered to produce only enough to cover everyone's needs. A bit more cautiously, Dumont (1973: 19) prefaced his self-described "utopian" plans with the claim that their purpose was to elicit constructive criticism, not to serve as blueprints. Still, like Gorz, Dumont (1977: 277) saw no profound tension between calls for "planetary organization" and centralized resource distribution, on the one hand, and "a decentralized and self-managed society" on the other.

The problem with ecosocialist utopian speculation is twofold. First, it attributes to central planning a degree of efficacy and beneficence that it has rarely if ever achieved in practice. On this issue, the anti-statist personalists have been the more prescient thinkers. Not until 1989 did Dumont clearly admit that the idea of collectivizing the principal means of production and exchange had "obviously failed on the economic level" (1989: 138). Still a ferocious critic of economic liberalism, Dumont now sees a Third Worldist ecosocialism as a matter of using incentives, credits, taxes, and innovative negotiated agreements to achieve egalitarian objectives. The second aspect of the problem is that ecosocialist utopianism's comprehensive vision of new patterns of production and exchange rests on a vastly oversimplified understanding of the values people pursue in their social relations. It is crucial to bring to people's attention the fact that some of their practices have potentially dire consequences for earth's ecosystems. But using theory to override their attachments to competing values (e.g., aesthetic diversity and privacy) does them no respect. Gorz continues to believe that the best "lever" with which to move society toward ecologically necessary reductions in consumption is a politics based on his autogestionnaire ideas. He now advances those ideas, however, not in a utopian vision but in a proposed "social contract" worked out in forums in which unions, consumer groups, the unemployed, and other relevant social groups represent their own views (Gorz 1991: 185, 206).

The notion of a contractualist ecosocialism, which Gorz mentions only casually, is central to the work of Jean-Paul Deléage and Alain Lipietz. Alternatively, ecosocialism can attempt to favor pluralism and decentralization not by appreciating the diversity of values but by relativizing them. The pitfalls of that approach are evident in the French ecosocialism of the 1980s.

Ecosocialism Relativized

In the 1980s, French ecosocialism was swept into a broader "Arc en Ciel" movement—a "rainbow" of leftists who could abide neither George Marchais's obstinately unreformed Communist Party nor François Mitterrand's comfortably reformist Socialist Party. Under Antoine Waechter's influence, the newly consolidated green party, Les Verts, was more interested in asserting a distinctive ecological identity than in creating broad alliances. With the 1988 presidential elections approaching, intellectuals and activists gathered more than 1,000 signatures in an attempt to overcome divisions between the autogestionnaire and ecological movements (Bennahmias and Roche 1992: 79)

Supporting the "Appel pour un arc en ciel" were two historical figures of French and German ecologism: René Dumont and Daniel Cohn-Bendit.

Joining them were the left-leaning Verts Félix Guattari and Dominique Voynet. And there were some supporters (e.g., the economist Alain Lipietz) who had long records of militancy in favor of an alternative left. This attempt to constitute a heterogeneous leftist coalition, although short-lived, established the distinctive tone of French ecosocialist writings in the 1980s. It relayed the autogestionnaire and Third Worldist themes of Gorz and Dumont without managing to mitigate the tensions between the centralizing and decentralizing interpretations of ecosocialism that run through their works. The essays of Pierre Juquin and Félix Guattari transcribe this irresolution into theory.

As a spokesman for the French Communist Party, Pierre Juquin was in charge of the party's positions on ecological issues in the 1970s. He had long been something of a maverick by the standards of Marchais's ossified organization. Encouraged by the enthusiasm for the Arc en ciel movement, Juquin announced his intention to run a Red and Green ticket in the 1988 presidential elections. Exclusion from the Communist Party followed forthwith. In his campaign, Juquin sought to unite all those who wanted to renovate the French far left. After obtaining only extremely modest electoral results, he then tried to establish a "Nouvelle Gauche" movement—again with little success (Jacob 1999: 277). His alternatives exhausted, Juquin applied to join Les Verts. He was admitted in October 1991 after rigorous questioning by party members who feared that their green waters were about to be invaded by "red submarines" (*Le Quotidien*, August 30, 1991). Had they read the manifesto that Juquin and others had published in 1990, they would have discovered a work more unsettling for its philosophical slipperiness than for any covert program of ideological aggression.

Explicitly socialist in their leanings, Juquin and the other authors of *Pour une alternative verte en Europe* put ecological disruption on the list of evils inherent in an economy organized to maximize profit, along with poverty, exploitation of workers, and imperialism. They warn against "eco-Keynesian" approaches to the environment on the ground that such approaches leave untouched the essential factors of disorder in the economy. The dominance of exchange value under capitalism undermines attempts to give value to ecological goods.

Although Marx is the only thinker whom Juquin et al. consistently cite favorably, they also note his shortcomings: His notion that the state of human consciousness is tightly linked to society's material processes of production smacks of "economistic" thinking. His vision of the proletariat as the exclusive agent of universal emancipation fails to allow for the diversity of cultures in the world, or other forms of "collective consciousness" that emerge when people become aware of their shared identity as members of a gender or as members of a species. Juquin and his colleagues pin their emancipatory hopes on various contemporary movements, including ecologism, feminism, multiculturalism, and pacifism. To correct Marx's shortcomings, they argue, the "green alternative" must be self-consciously "relativistic" (Juquin et al. 1990: 43–44).

As theory, this position is weakened by a confusion between ethical pluralism and ethical relativism. Ethical relativists hold that moral beliefs are entirely dependent on context; there are no universal values. Ethical pluralists hold that moral beliefs are complex; they consist of a great variety of claims about the expression of values. Many of those claims *are* universal, but they also conflict with one another, and there is no common metric that might settle conflicts by definitively ranking values. The difference between the two positions is subtle but crucial. Relativism easily leads to the conclusion that there are no grounds for argument between defenders of conflicting moral positions. Some love undisturbed nature; others prefer a lifestyle of material gratifications and bracing competition. Since each person's view depends on context, there is no hope of leading them to a reasoned reconciliation. That is the consequence—probably unintentional—of Juquin's aspiration to "relativize" Marx. I say "unintentional" because it appears that Juquin does not really mean to abandon a standpoint in which some choices truly *are* superior to others.

What Juquin really seems to favor is revealed when he demands "a decentralized, participatory, and as much as possible, direct democracy" (Juquin et al. 1990: 50). Like an ethical pluralist, he calls for forums in which people can articulate their viewpoints and adjudicate their differences. The resulting compromises do not testify to the relativity of values. Where values are multiple and irreducible, compromise is not tantamount to betrayal of principle. To compromise is to respect others and to recognize the *moral* importance of social cooperation (Lukes 1991: 226). That is a line of reasoning that leads toward the social contract and, I will argue, toward Habermas's account of a discursively grounded consensus.

To be unswervingly relativistic, an ecosocialist would have to question the possibility or desirability of contractual agreement itself. That path was explored by Félix Guattari. Guattari turned difference—"singularization," he called it—into the leitmotif of leftist ecologism. Guattari had been a Trotskyite and a Communist. (He was suspended from the French Communist Party in 1956.) He underwent psychoanalysis with Jacques Lacan in the 1960s and joined the Freudian School. Then, repelled by the authoritarian rigidities of Freudian practice, he became a major figure at the psychiatric clinic at La Borde—a haven for radicals whose sparring with "bourgeois" society had left them needing sympathetic counseling. Before his death in 1992, Guattari worked closely with those who were socially disaffected or politically marginalized. He formed groups promoting radical politics and progressive thought in a variety of issue areas, including psychiatric reform, Latin American revolution, urban planning, independent radio broadcasting, and homosexual rights (*Libération*, August 31, 1992). Disappointment with the official left in France led him to conclude that "either socialism will come to focus its concern on changing ordinary life, on human relations of proximity and solidarity . . . or it will disappear from the map of hope and perhaps make way for a new ecological pole" (*Le Monde*, July 12, 1990). His last engagement on behalf of such a "molecular revolution" consisted of joining both Les Verts and its rival, Brice Lalonde's Génération Écologie, and seeking to unify them.

Guattari's ecosocialism draws its categories of analysis from his earlier work with Gilles Deleuze. Their joint work *L'Anti-Oedipe* (Deleuze and Guattari 1972) stresses the revolutionizing, creative potential of unbounded desire. In that work, Deleuze and Guattari posit that desire in the individual is neither an effect of society's relations of production nor a primitive, libidinous force that determines consciousness. They picture desire as a surging, productive energy. Like Nietzsche's will to power, it is capable of creativity and self-affirmation, but it is also subject to paranoiac impulses to form repressive social institutions that thrive on life-denying repetition.

Capitalism, Deleuze and Guattari argue, is the historical culmination of the latter tendency. Capitalism "deterritorializes" the representational systems of primitive and despotic societies. It breaks down all traditional social codes, including kinship, class systems, religious beliefs, and folk traditions. Freed from social codes, all goods—ephemeral images as well as durable products, labor and knowledge as well as consumable commodities—are allowed to find their equivalents on the market. But this decoding does not liberate the creative, individuating potential of desire. It only prepares the ground for *reterritorializing* the psyche. According to Deleuze and Guattari (1972: 277), "all the decoded flows, including the flows of scientific and technical code," are organized "for the benefit of the capitalist system and in the service of its ends." Interpreting individuals merely as "abstract quantities" (e.g., labor and capital), capitalism normalizes them for the purpose of furthering capital accumulation. The only hope for escaping this system lies in multiplying the number, the activities, and the interconnections of groups that eschew internal hierarchies and whose members are not responsive to the constraints of commodity exchange (Bogue 1989: 103). That is essentially the agenda of Guattari's late "ecosophical" writings, *Les Trois Écologies* and *Chaosmose*, in which he views "ecology" expansively as including any mode of thinking that concerns itself with humankind's efforts to "make the world habitable." Those efforts take place at three levels.

"Mental ecology" deals with the creation of meaning, with the ways that individual subjectivity takes form under the influence of socio-economic ensembles. "Integrated World Capitalism," Guattari charges, disrupts the sources of aesthetic and ethical creativity by infiltrating its homogenizing norms into every aspect of everyday life. The ecosophical approach, in contrast, promotes "innovative practices" and "the proliferation of alternative experiments that focus on respect for singularity and on a permanent effort to produce subjectivity" (Guattari 1989: 57).

"Social ecology" deals with the sources of solidarity that make social life pragmatically effective and emotionally satisfying. Capitalism destroys social bonds, driving successful individuals to compete in divisively hierarchical organizations and to live in soulless suburbs, leaving the unsuccessful behind, jobless, in menacing urban ghettos. The sedating power of the media facilitates this entire process. The ecosophical strategy is to "reappropriate the media by a multitude of group-subjects which are capable of guiding them on the path of resingularization" (ibid.: 61).

"Environmental ecology" deals with the biospheric equilibria upon which all earthly life depends. Integrated World Capitalism achieves economic growth only at the price of ecological devastation and social segre gation. The ecosophical response is a federalist redefinition of the state's multiple functions, the recentering of economic activity on the production of subjectivity, and the building of an ecology movement that focuses more on its mental and social ecology than on preserving nature per se (Guattari 1992: 169).

From one perspective, Guattari's ecosophy is quintessential French ecologism. It offers an ecologism that is noncentered because it advocates an ongoing reassessment of "nature," particularly in relation to the often-repressed possibilities of human subjectivity (Conley 1997: 93–99). Using postmodern tools, ecosophy relativizes Marx's materialist theory of ideological production. In a theory of dialectical historical movement, the self-centered, calculating, consuming individual of liberal politics is a mode of subjectivity generated by a particular way of organizing material production. Guattari cannot accept such a theory to the extent that it posits a linear causality between an economic base and an ideological superstructure, so his "chaosmotic" approach to ecologism tries to follow "how various changes in the use of machines and technologies [*mutations machiniques*] contaminate and influence one another, how they create loci of subjectivity" (Guattari 1991a). Precisely because societies and machines form unique and fluid ensembles, it is impossible to speak of the subjects within them as centered.

From another perspective, however, Guattari's postmodern moves cause him to stretch the strands of noncentered ecologism past the breaking point. In effect, he completely relativizes ecology to the human capacity for expressive subjectivity. He believes that the key to avoiding disasters in the biosphere is propagating everything about individuals that is singular, emotive, "fantasmic," inventive, and value creating. Arguably, mobilizing groups to fight stereotypes.propagated in the capitalist media or to favor miniaturized technologies may have effects on wetlands or oil-drilling practices (Conley 1997: 96, 97), but Guattari leaves it to sympathetic interpreters to make that connection. He never actually defends his "ecosophy" against an obvious objection. Recognizing ecological duties could easily require curtailing at least some forms of creative effusion. His call to "respect the heterogeneity of values" sounds like ethical pluralism—except that it has great difficulty appreciating the ethical value of consensual processes that might mitigate conflict. In dealing with ecosophy's opponents, Guattari

(1991b) suggests maintaining a "dissensual" relationship of ongoing polemics and efforts to see the other's view of the world.

It is easy to see why Guattari cannot use the language of consensus and contract: As soon as one compromises in order to attain a degree of group solidarity, one is by definition less "singular" than before. Dissensus sounds like a perfect formula for perpetuating the divisions within the ecology movement that have so often stymied its efforts to achieve political efficacy. And there is an additional irony: For all Guattari's talk of difference, his view of Integrated World Capitalism is stunningly devoid of nuance. There is no hint that different "capitalist" countries have achieved different mixes of markets, welfare-state benefits, and environmental regulation. Then again, Guattari's indifference to such differences is not so surprising after all. The multiple forms of capitalism have resulted from consensual processes that do not fit into Guattari's framework. Relativist ecosocialism ends in a philosophical impasse.

Contractual Ecosocialism I: Value and Scientific Ecology

With two more recent converts—Jean-Paul Deléage and Alain Lipietz—French ecosocialism may have cleared many of the obstructions from its path. Deléage and Lipietz bring to ecosocialism a greater appreciation of its scientific foundations and political presuppositions. They carry on Dumont's Third Worldism, Gorz's critique of economic rationality, and Juquin's and Guattari's demands for a more participatory ecologism. And, like their predecessors, they understand that their goals do not form a single, neat, internally consistent social project informed by a complete ethical theory. Their contribution is to take these ideas and lessons one step further by explicitly interpreting ecosocialism contractually. Contractual ecosocialists regard ideals of equality and autonomy as the fundamental values of ecological negotiators who seek to win the assent of diverse groups to a stable and environmentally sustainable social order.

Jean-Paul Deléage is a professor of physics and the history of science at the Université de Paris VII. From 1970 to 1976, as a member of the Trotskyite Ligue Communiste révolutionnaire, he was active in campaigns to protect workers against asbestos. Long stays in Tunisia and Nigeria sensitized him to the need for an ecological politics that does not shortchange the interests of less developed countries.[4] After breaking with the Trotskyites, he devoted himself to the study of scientific ecology. In 1986 he co-authored a work arguing that a society's successes or failures in adapting its "energy systems" to changing environmental and social conditions help determine its class structure, its material prosperity, and its political stability (Debeir, Deléage, and Hémery 1986). Deléage reentered politics at the end of the 1980s, when he directed Pierre Juquin's presidential campaign. He joined Les Verts in 1989, and by late 1991 he was on that party's executive committee (Pronier and Seigneur 1992: 240–242).

Deléage's background makes him an especially acute commentator on the relationship between scientific ecology and social structures. His major work to date, *Histoire de l'écologie*, traces out multiple, reciprocally conditioning factors in the relationship between societies and nature. Deléage presupposes no unified theory connecting social and natural evolution. Indeed, he discounts the possibility of such a theory: "The categories of ecology, even if it is called human ecology, cannot by themselves account for the exchanges between man and nature. The modalities of these exchanges evolve along with social structures, and the latter resist ecological analysis." (Deléage 1992b: 245) Ecologized accounts of social change will be ones that are attentive to the ways in which environmental transformations are regularly factored into explanations of the flourishing and decay of human communities. Deléage investigates the impact of environmental change on civilizations as diverse as ancient Mesopotamia and sixteenth-century England. Societies have adapted, or failed to adapt, to droughts, to degradation of forests, to plant diseases, and to the silting up of rivers. Some of these problems are caused or worsened by human activities; others are simply natural misfortunes. How significant these environmental changes are for social development depends, unpredictably, on the society's ability to devise compensatory technologies and more supple forms of organization.

Not only social structures but also social "representations"—systems of ideas, world views, ideologies—shape humans' dealings with the environment. Cosmologies and myths that sacralized nature limited the ways in which societies of pre-industrial Europe's exploited their environment. Similarly, the tendency of today's capitalist societies to conceive of value primarily in monetary terms is only a time-bound social representation.

Borrowing terms from Gorz, Deléage notes that "particular meanings" have enriched societies less dominated by cycles of production and consumption. Today, feelings of class solidarity, feelings of attachment to a place, feelings of rapport with nature, religious beliefs, and philosophical convictions all fall victim to "a logic of universal economic exchange" (Deléage 1993b: 9). That same logic accelerates the pace at which natural resources are transformed into exchangeable goods (Deléage 1992b: 265, 295). Deléage makes no attempt to establish the direction of causation between social structures and discourses. It is enough that the two are reciprocally influential.

Throughout his writings, Deléage emphasizes the global dimensions of environmental change in the contemporary world. He insists that the history of ecology reveals the development of "world ecology," insofar as "the creation of a world productive space has laid the basis for the ecological unification of the world" (Deléage and Hémery 1990: 30). Ecologically destabilizing practices (industrialization, high consumption of fossil fuels, genetic uniformity in agriculturally exploited species, destruction of forests, discharges of dangerous chemicals into air and water), combined with the burgeoning of the human population, have intensified to the point where they affect homeostatic processes throughout the biosphere. This implies that we must "enlarge still more the field of ecological reflection" so that we include human activity as *a part of* ecology's domain. Since human activity depends on how people are organized in communities and on how they conceive their goals, matters heretofore left to various disciplines of the social sciences must be included in our ecological understanding. "Ecology," concludes Deléage (1992b: 297), "is necessarily polydisciplinary, situated at the crossroads of natural sciences like biology and earth sciences and the human sciences like ethnology or economics."

Scientific ecologists may recoil from so vast an intellectual mandate. Some will object that bringing the human sciences into the domain of ecology risks compromising the objectivity of ecological studies. Deléage agrees in part: He notes that "it is often difficult to eliminate all value judgments from the objects of [ecological] study," and that in a multidisciplinary ecology it will be all the more "difficult to eliminate the observer's viewpoint from which the living reality is perceived" (Deléage 1993c: 130–131). But part of his point is that these value judgments are already being made.

Decisions about what ecosystems to study or about the scope of what constitutes a system are already shaped by value-laden social processes. Deciding whether to include human activity within ecosystems is not a value-free choice. There is, for example, a link between a scientific ecology that limits itself to the study of ecosystems defined independent of human activity and an "ecomanagement" approach to environmental politics (Deléage 1993c: 128). The narrow view will tend to pick out relatively pristine areas for special protection. The increasingly rare ecosystem in which the human impact is minimal will be valorized by a science that has chosen such systems as its special object of study. When society decides to protect those areas, it will need people whose expertise allows them to track and protect the delicate equilibria that characterize that particular ecosystem. Meanwhile, the human activities that have made such ecosystems so rare will go unquestioned. Ecosystems already altered by man may be abandoned to complete destruction.

Moreover, an ecology that sees human activity as external to natural processes may simply be used to make industrial or agricultural production more efficient. If crop irrigation in a region is depleting the groundwater, ecologically trained agronomists and hydrologists may be able to recommend changes in irrigation techniques—changes that protect the water supply while making production more secure for the long term. Yet such recommendations have a distinctly conservative cast (Deléage 1992b: 301). They fail to ask certain questions: Is this crop really appropriate to this region? Do other problems (e.g., excessive consumption of fuel, soil erosion, reductions in genetic diversity, disruption of animal migratory patterns) argue against this sort of use of the land in this region? Is a profit-hungry agribusiness simply going to create a new set of environmental problems once the current one is brought under control? What social structures are creating demand for this sort of agricultural production? Are those structures capable of functioning within a system of distribution that meets the needs of all people? A refusal to ask these sort of questions makes the narrow interpretation of ecology the accomplice of practices that perpetuate environmental destruction and political exclusion.

Deléage's enlarged ecological reflection shows that ecosocialists need not relativize "nature" to the perceptions of the desiring individual or to the demands of politicized groups, as if science had no special standing in the

understanding of ecological processes. His point is that even the most sophisticated, scientific understandings of natural phenomena take shape in a
framework of social priorities and prejudices. No one has access to a totalizing perspective that might justify automatic deference on the part of ordinary citizens. There are only "partial objectivities" that flow out of social
groups with diverse economic, sexual, and cultural orientations (Deléage
1992a: 9). Their voices have a significant role to play in correcting the distortions induced by the social context of science. Deléage looks forward to
"a culture of counter-expertise" in which scientific analysis would have to
answer to more probing ethical scrutiny. At the same time, he regards science
as indispensable in helping citizens understand the ecological consequences
of their actions and the alternatives—the margin of freedom—left to them.
Ecosocialism should be critically rational, closely tied to diverse social
groups, and thoroughly democratic. As Deléage recognizes, this vision of a
pluralistic ecosocialism implies "reconceptualizing not only mankind's
belonging to nature, but also the social contract" (1993b: 12–14). How such
a new contractualism might integrate material productivity, social equality,
and environmental responsibility Deléage leaves as an open challenge to
theoretical reflection. Alain Lipietz takes up this challenge.

Contractual Ecosocialism II: Value and Economic Regulation

Affiliated since the 1970s with a research institute concerned with economic
planning, Alain Lipietz is known as a leader in the "regulation school" of
political economic theory in France.[5] After years of participating in political movements to the left of François Mitterrand's Socialist Party, Lipietz
decided to join Les Verts in 1988. Within only a few years he was named
spokesman of the party's economic commission and elected a regional coun-
cillor in the Paris region (Gher 1992). He was the main author of Les Verts'
1992 economic program. In June 2001 he was selected as the party's presidential candidate for the 2002 national elections.

In *Vert Espérance* [*Green Hope*] (1993a), Lipietz recounts the intellectual
changes that led him from expounding an ideological mélange of communism and French Maoism to embracing political ecology. Today he no
longer believes in the centrality of the workers' movement; he no longer
believes that capitalism is the unique source of all forms of oppression. He

now strives to incorporate insights drawn from diverse oppressed groups (women, gays, peoples of the Third World) and from ecologists into his understanding of the opportunities for social transformation. Yet Lipietz also sees his evolution in terms of a core commitment that he has simply extended through time. Its unifying theme is a "revolt against an unjust economic order, which tears society apart into rich and poor, which sullies nature because it does not even respect human dignity" (Lipietz 1993a: 7)

Lipietz's Red-green perspective places him among the more Marxist adherents of "regulationism," a post-Keynesian school of economic thought that has flourished in France since the 1970s. Robert Boyer, one of the main proponents of regulation theory, explains that this school's approach originates in a rejection of the methodological individualism of mainstream economics. Methodological individualists explain social phenomena in terms of the behavior of strategically rational individuals, irrespective of their place in social categories such as class, race, and gender. For example, in explaining where individuals end up in the social division of labor, neoclassical economists may look at the individuals' own choices of education or of profession and at the demand for certain products constituted by other individuals bidding in the market. In contrast, regulationists show how socially structured patterns of behavior (e.g., wage and commodity relations) themselves guide individual choices. Lipietz, who ponders the theory's metaphysical underpinnings more than some of his *confrères* do, begins explicitly from the materialist view that society—particularly through need-oriented production and exchange—fashions the motives of individuals. At the same time, needs themselves evolve in the context of struggles between social groups, on the basis of limited resources passed on by previous generations (Lipietz 1988b: 12–14).

The social structures of the greatest importance to regulationists are those that make possible the growth of productive capital. Regulation theory describes an economy oriented not toward "general equilibrium" but toward "phases of expansion and moderate cyclical fluctuations, followed by phases of stagnation and instability" (Boyer 1990: 13). Capital accumulation is thus not a smooth, self-governing process; it is beset with recurring crises of overproduction, unemployment, and social turmoil. It is the stability and reproduction of socio-economic systems, not crisis as such, that most urgently require explanation. Regulationists hypothesize that a "mode of regulation" mitigates disorder in a "regime of accumulation" (Lipietz 1992b: 103). Lipietz adds that a particular "model of work organization" governing the division of labor and structures of authority within firms forms an integral part of a stable "model of development."

According to Lipietz (1989: 16), to study a "regime of accumulation" is to examine at a macroeconomic level how production (mechanization, importance of various sectors of the economy, worker productivity) and use of the social product (for personal consumption, investment, trade, etc.) evolve together and support each other. In view of the conflictual nature of capitalist development, however, a regime's longevity depends on a "mode of regulation" to become stable. The French sense of the word "regulation" goes far beyond the American sense (i.e., government intervention to correct potential market failures or to control monopolies). The French sense designates a variety of social mechanisms that attenuate conflicts within a set of social relations, allowing those relations to reproduce. A mode of regulation includes behavioral norms (e.g., seeing certain forms of workplace hierarchy as legitimate), welfare legislation, union contracts, and state-mandated safety regulations.

Under what conditions are such regulatory arrangements created? Conflict is transformed—always temporarily—into social reproduction when competing groups arrive at a set of compromises over how to organize the production and distribution of social goods (Lipietz 1999: 87). In their struggle for advantage, groups eventually press one another to accept limits, rules, procedures, divisions of territory, rights, and duties. Social mobilization and negotiated settlements, backed up by state-sanctioned rules, help steady the regime. With the system's contradictions temporarily tamed, capital accumulation then proceeds apace until new crises force further adjustments.

According to this model of social explanation, stability is won through the creation of "hegemonic historical blocs" (Lipietz 1988a: 82). A regime of accumulation typically gives disproportionate advantages to certain groups. Yet widespread voluntary acceptance of its institutions and norms is crucial to its stability. Lipietz draws on Pierre Bourdieu's notion of *habitus* to explain how a regime fosters "appropriate" individual expectations about work, consumption, and life chances. These dispositions help to fit most individuals smoothly into social roles functional to the regime (Lipietz

1987: 18). The state furthers this process of normalization by putting its legitimizing imprimatur on the compromises and customs that form the hegemonic system.

Still, nothing guarantees the long-term success of such efforts. A regime of accumulation may eventually be unable to fulfill all the expectations it creates. Changes in technology, in trade, or in available resources may cause unforeseen friction among the pieces of the hegemonic system. Indeed, this is to be expected, since regulation lessens social tensions but does not eliminate them—at best, it creates "armistices" within class struggles. Capitalist "extortion of surplus value" remains (Lipietz 1994: 41). A crisis occurs when the system of regulation shows itself unable to stem mounting productivity losses, trade deficits, and sociopolitical turmoil. Social actors then search for the terms of a new compromise—one better able to manage the accumulated tensions of the previous regime.

Most regulationists apply this general schema for understanding social change to the postwar political economy of Europe and the United States. Lipietz especially contends that the "Fordist" social compromises that underwrote prosperity in the 20 years that followed World War II have been unraveling since the late 1960s. In his 1989 book *Choisir l'audace* [*Choosing Boldly*, published in English under the title *Towards a New Economic Order*] he explains the nature of the crisis and, for the first time in his major writings, links it to ecological concerns.

Lipietz argues that what sustained the relative social peace and economic resurgence of most industrialized Western countries after World War II was a "Fordist" regime of accumulation. Fordism couples a model of work organization based on high levels of mechanization and Taylorist "rationalization"[6] with agreements to distribute the fruits of economic growth widely within the nation. The first element of Fordism disadvantaged workers by devaluing the knowledge of the production process they had gained on the shop floor, thereby making work less fulfilling. It made workers easier to replace, thus potentially lowering wages. The resulting possibilities of labor unrest and declining productivity made a second element of Fordism essential: In compensation for their diminished position in the workplace, workers demanded that management redistribute more of the profits to them.[7] In the 1930s and the 1940s, from the American New Deal to the Scandinavian social democracies, to the French model of economic planning, governments oversaw the compromises that attenuated tensions between capital and labor. By combining free enterprise with union contracts, extensive government regulation and welfare policies, a temporarily workable compromise was struck.

The eventual breakdown of this agreement follows, in part, from its own internal logic. The system unravels because its field of incentives generates long-term behavioral consequences contrary to its own premises. Increasing mechanization and computerization and subcontracting of manual labor to areas where wages are low, says Lipietz, are strategies that only exacerbate Taylorism: Workers whose knowledge and talents are excluded from their firms' organizational plans become less productive. Diminished profit brings diminished investment, and eventually it causes unemployment and reduces tax revenues for the welfare state. Meanwhile, the growing internationalization of trade worsens the crisis. Heightened competition among the United States, Europe, and Japan brings calls to roll back regulation at the national level. Reducing the state's economic intervention boosts production only by exacerbating trends toward social inequality. Lipietz demonstrates how current economic trends undermine the mode of regulation that made the Fordist regime of accumulation work.

Though this understanding of economic crisis is common among regulationists, it takes an ecological turn when Lipietz introduces a new category of analysis into it. Since 1988 Lipietz has argued not only that the various compromises are generating self-defeating behavioral effects but also that they rely on a strategy of social stabilization that is ultimately incompatible with ecologically sustainable economic development. His warning that "the era of the finite earth has begun" is aimed as much at regulationists as at liberals. Taking account of this new factor requires regulation theorists to analyze a previously unrecognized premise of twentieth-century regimes of accumulation: their underlying productivist strategy.

Capitalism always tends toward productivism. Growth is inherent in its logic. One invests in capital for profit, yet sustaining profit requires expanding market share. Firms that fail to grow must eventually see their products superseded by competitors bent on capturing their profits for themselves. Nonetheless, capitalism's growth tendency was held in check until the middle of the twentieth century by the system's own distributive and organizational weaknesses. It concentrated wealth in the hands of a relatively few capitalists, whose ability to consume was necessarily limited. This created crises of overproduction, periodically disrupting the growth patterns of the economy. Inefficient, pre-Taylorist organization of work processes slowed increases in productivity.

Although the Fordist compromise admitted some collectivist constraints on economic activity (e.g., channeling some profits into pensions or safety regulations), it did not really challenge productivism. Instead, it perfected it. Giving workers additional income and security, it turned them into consumers capable of absorbing the increased output of scientifically rationalized work processes. Thus, the Fordist compromise removed earlier impediments to economic expansion. Not only *could* growth accelerate; it *had* to. "Free enterprise" became responsible for generating sufficient capital to fund not only accumulation and profit but also higher wages and some of the charges of the welfare state. More than ever, the social logic of productivism prevailed. The stability of such arrangements depended on one key assumption: that growth must continue unabated.

The connection with ecological concerns is made when one realizes that the Fordist compromise involved trading away the livability of the environment. Its productivist premise implicitly denied the finitude of the planet's capacity to supply the raw materials of production and to absorb its waste products. This first became apparent when firms ended up casting off so much waste or so intensively exploiting resources that enterprises begin to interfere with one another's profitability. The greenhouse effect and the pollution of the oceans raised popular awareness that current rates of production and consumption undermine the well-being of future generations (Lipietz 1989: 62–64). Yet the companies obliged to clean up their waste complain that such efforts force them to raise prices, diminish production, and cut back employment.—steps that further weaken the foundation of a regime of accumulation whose stability depends on maximizing production and consumption. Only a new social compromise, Lipietz argues, can resolve the crisis.

Obviously, not just any social compromise will do. To mitigate the sources of social instability arising from the exhaustion of Fordism *and* to respect the ecological constraints facing humanity, the new compromise must embody a nonproductivist set of values that can, arguably, settle into a self-reinforcing system. Lipietz (1993a: 18–19) identifies those values as solidarity, autonomy, ecological responsibility, and democracy. If the post-war compromise is exhausted in part because technological change has reduced the demand for labor, then a new compromise would combine solidarity and ecological responsibility by more equitably distributing work and free time. If the Fordist regime of accumulation allowed private enterprise to "socialize" the costs of environmental damage in the form of pollution and destruction of the landscape, then an ecologically responsible compromise must use state taxes, subsidies, and development strategies to restore and protect the environment. Such means, however, usually imply transferring even more power to a bureaucratized, centralized state, thereby weakening the value of democracy. To express democracy and ecological responsibility simultaneously, Lipietz proposes fostering political activism by progressive grassroots organizations. State intervention can be avoided if society is composed of organized interest groups that express their conflicting interests in face-to-face dialogue, arriving at tolerable compromises (ibid.: 28–29; 1989: 70–71).[8]

Simultaneously, to counter the destabilizing social and ecological effects of internationally mobile capital, institutions with transnational regulatory powers are needed. Social compromises—this time in the form of international agreements—are necessary to prevent countries' internal compromises from being undermined by competition to retain or attract investment (Lipietz 1993a: 62, 80; 1999: 120–123). Owing much to René Dumont, Lipietz is a Third Worldist who proposes that the North transfer needed funds and technologies to the South to promote sustainable development. Through negotiations, North and South should seek a "minimal universalism" to "unify progressively social and ecological norms" throughout the world (Lipietz 1993a: 80).

The Ethics of the Green Contract

What does it mean to regard these proposals as a "compromise" rather than as a utopian vision or as subjective "resingularization"? It means, first, seeing society as composed of groups with widely divergent values. Contemporary ecosocialism parts with any Marxian hopes that a particular class is destined to fulfill mankind's quest for mutual recognition and social reconciliation. Like the later Gorz, Lipietz understands that moving beyond conflict requires mutually respecting discussion, not utopian visions of social harmony. Second, contractual ecosocialism advances its program through a broadly participatory, *consensus-oriented* process of negotiation. Although Lipietz admires Guattari, he does not simply praise dissensus. For Lipietz (1988a: 84), the Alternative "defines itself in terms of a community sharing a minimal agreement of what is just and good to build together." Greens "seek to convince others, by the rightness of their proposals and the example of their work, to come together in a common struggle for life" (Lipietz 1993a: 41).

Yet moving toward what is "just and good" may require painful sacrifices on the part of those who have benefited from unjust practices. A contractualist view therefore requires putting together a package of changes that includes measures designed to secure the consent of potentially recalcitrant groups. Along this line, Lipietz (1993a: 84) argues that ending the exploitation of the peoples of the South will never win majority support in the North unless it is tied to ideas such as expanding free time and supporting "more festive" forms of social life. In a sense, then, he tries to create an attractive formula out of the partialities of Dumont's Third Worldism and Gorz's autogestionnaire agenda. The former has the politically unpalatable consequence of defining ecosocialism largely in terms of austerity. (See, e.g., Dumont 1978: 184.) The latter has the ethically problematic characteristic of neglecting the specificity of Third World development. Joined together as parts of a social compromise, however, these two currents help remedy each other's defects.

Thus, the ecological turn in contractualist theorizing brings together the regulationist emphasis on the consensual resolution of social conflict with an environmentalist value system—one that defends ecologically sustainable development and respect for nature. Yet it is by no means clear how ecosocialist contractualism melds explanation and evaluation. Can one derive green values through the same cognitive processes of observation, reasoning, and testing that support regulation theory? If so, how? If not, with what sort of argument does one support green values? How do the motivations imputed to social actors in regulation theory tally with the transformed ethics presupposed by a green society? What is the relationship between "compromise" as a variable in regulation theory's explanation of social reproduction and as a legitimizing ethical conception ? Since Lipietz draws on contemporary social theory more than any other French ecosocialist, it is worth probing his writings further for answers to such questions.

In point of fact, Lipietz rarely tackles ethical questions directly. But when he does, he sounds distinctly Weberian. Regulation theory, he implies, yields explanations that are analytically distinct from the theorist's own ethical convictions. As social theorists, Lipietz says, the task of regulationists is to search for relatively "fixed tracks" in the midst of conflictual group relations. Various social compromises create a relatively enduring system, allowing the theorist to examine the functional interdependence of the ways of organizing work, regulatory regimes, state economic and welfare policies, popular values, and so forth.

Evaluation is another matter. "One can have an ethical judgment about a form of social relations . . . ," remarks Lipietz (1987: 22), "but no one can say there is something like historical progress. . . ." Apparently, whatever moral judgments the theorist might have about the quality of the compromises—whether they are distributionally fair, whether they come at too great an expense in the destruction of nature—originate in an ethics devised independent of the process of social change. The theorist can propose ethically superior arrangements. But the ethics are his, not history's. Alternative consensual arrangements will "win" only if taken up by groups willing to struggle for them. Deliberately or not, Lipietz follows Weber's (1949: 11) injunction that "the investigator . . . should keep unconditionally separate the establishment of empirical facts . . . and his own practical evaluations, i.e., his evaluation of those facts as satisfactory or unsatisfactory." This separation of social and ethical theory does justice to neither. At the ethical level, this interpretation runs into the same danger as Guattari's relativism. Liberal productivists have their values; Lipietz and the new social movements put forth different ones. Neither makes its case in terms of superior rationality. The two simply clash in struggle and compromise.

But that is not really the form of Lipietz's own arguments for an ecological politics. Throughout *Vert Espérance* Lipietz makes judgments that presuppose the comparability of different values—and the superiority of green ones. When he characterizes biological diversity as the "immune system of our biosphere," he chooses an image designed to make us all see the folly of wrecking the very system that supports our lives. He denounces distributive systems that allow the wealthy North to capture a disproportionate share of the planet's resources, disregarding the more urgent needs of the South. He assesses the relative ability of different strategies of economic development to improve the quality of life of all. In all such cases, Lipietz is proposing more than an alternative ethic which others may or may not find attractive. He lays out a candidate for a system of values that is more consistent, that better meets our own stated goals, that does not make unwarranted assumptions—in a word, that is more *rational*, in a sense not conveyed in his more Weberian statements.

Equally unsatisfying, Lipietz disconnects his "green" values from the nature of the crisis itself by making it seem that the values of an ecological ethics arise in ways that are entirely distinct from cognitive processes of observation, reasoning, and testing. Ecological challenges do not articulate with human interests in any systematic way, such that one might *expect* certain conditions of environmental change to favor the development of a "green" consciousness. Demands to protect biological diversity, for example, appear to arise simply because some new social movements decide that such protection is important—not because at this particular historical juncture humankind's species-depleting activities have reached a point where they are particularly likely to activate an interest in the preservation of ecological systems. A decisionist model of moral criticism deflects theorists from the important task of explaining the origin of interests or of the conditions favoring their critical reevaluation.

In chapter 8, I will explore whether some of these unresolved problems might be better addressed within the theoretical framework proposed by Jürgen Habermas. Seeing historical change in terms of a critical, reflexive learning process might better encompass the French ecosocialists' commitments to democracy, social equity, and environmental care than some of their own explanatory claims.

But it must be admitted that the noncenteredness of French ecologism challenges any efforts to advance that theoretical convergence. Some would say that Habermas can interpret the natural world only as an object to be technologically dominated by a separate subject (di Norcia 1974–75: 89). Lipietz (1993a: 49) insists, in contrast, that the 'environment' is not at the center of political ecology. Rather, there is a complex totality, structured as a triangle: the human species, its activity, and nature. Nature—threatened and transformed by man—is both the matrix and the basis of that activity. . . . The *target* of political ecology can be none other than that decisive mediation between humanity and its environment: the productive, transformative and consuming activity of humanity.

How Habermasian theory might absorb ecological viewpoints in which nature is both "the matrix and the basis" of human activity will receive further attention in chapter 8.

For the moment, what is clear is that Lipietz's position, and that of the other French ecosocialists, makes nonsense of liberal charges of the fundamental anti-humanism of political ecology. Accusations that greens value "nature" more than humanity have not the remotest applicability to French ecosocialists. The next chapter examines why France's liberals fail to understand that, as Jean-Paul Deléage claims (1993a), "ecology is the humanism of our time." French liberals make their charges of anti-humanism credible only by refusing to confront French ecologists on their own terms. The problem may be that the liberals are so wedded to the centeredness assumption that they hardly know where to begin a debate with greens who challenge it. Only indirectly do they ask whether noncentered theories can make sense of values of liberty and democracy in the "age of a finite earth."

7

Negotiating Nature: Liberalizing or Democratizing?

"As a political movement, ecology will not be democratic; as a democratic movement, it will have to give up on the mirage of engaging in grand politics. . . . Between barbarism and humanism it is now up to democratic ecology to decide." With that peremptory formulation, Luc Ferry (1992: 268) sums up the sentiments of numerous French liberals.[1] They detect in green political thought a willingness to negate human rights in the name of "nature" and a proclivity to adopt authoritarian means to implement totalizing solutions to social problems.[2] In the early 1990s, Ferry and others thought it timely to launch critiques because Les Verts were experiencing an electoral upswing that might soon make them a party of government. Temptations of their buying wholesale into a comprehensive ideology were likely to grow. It was urgent to reveal the flaws in political ecology's theoretical underpinnings and, from a liberal point of view, to lay out the alternatives starkly: barbarism or humanism, authoritarianism or "democratic ecology."

Questions about the relationship between green values and democracy are common in English-speaking ecologism too. Some theorists argue that green values and democratic procedures are analytically distinct, leaving democracy as, at best, a means to ecological ends. A more authoritarian politics appears inevitable to some who perceive the urgency of green ends but doubt the effectiveness of democratic means to reach them. Green democrats then respond by trying to demonstrate a more integral link between democracy and ecology. France, in contrast, has no prominent defenders of eco-authoritarianism, and most theorists avoid the nature/humanity dualism that fuels the debate over whether democracy serves green values.

The absence of an indigenous philosophical defense of nature's intrinsic value does not stop French liberals from detecting authoritarian inclinations in green thought. It only forces them to seek historical and circumstantial (rather than textual) grounds for their accusations. French liberals associate greens with anti-modern conservative thought—even to the point of portraying them as the ideological soulmates of neo-ruralist fascists. In this way, they provide evidence for their contention that ecologists, to be genuinely democratic, must be liberals. Even if liberals would amend the Cartesian project of "mastering nature," they insist that something close to Descartes's understanding of human subjectivity is a necessary component of any ecologism that respects human rights. Unlike the revised humanisms proffered by other French ecologists, liberal "democratic ecology" depends on conceiving the self as centered relative to an ontologically fixed nature. The centeredness assumption prevents liberals from engaging critically with French ecologists' theories of the reciprocally problematizing processes that engender understandings of nature and humanity as distinct entities in the first place. French ecologists do not subordinate self-governing citizens to natural values (as liberals charge) any more than they treat democracy only as a contingently suitable instrument for building an ecological society (as English-speaking ecologists tend to do). Noncentered ecologists look to essay new forms of popular participation *at the sites where the linkage between humanity and nature gets forged.* Were French liberals not so preoccupied with the danger of eco-authoritarianism, they might see that greens in their own linguistic community are trying to do something different from their English-speaking counterparts. They are trying not just to inject more popular participation into environmental policy making but also to imagine how to democratize the very processes out of which "nature" gets constituted.

Democratic Ecology: The Centered Debate

The "democracy" that liberals have in mind when they speak of "democratic ecology" consists of political processes and institutions—mainly of national government—that are designed to make law and policies generally reflect the will of the majority, within constraints imposed by certain constitutional rights. Liberal democracy is neutral with respect to every substantive policy end. "Liberal political theory," explains Mark Sagoff

(1988: 162), "cannot commit a democracy beforehand to adopt any general rule or principles that answer the moral questions that confront it." In order to express equal respect for all citizens, liberal democracy favors no particular conception of human perfection. In order to leave citizens free to choose how best to order their lives, it establishes no religion and it forbids no association that works within the constitutional order.

The supposition that democracy must be substantively neutral explains Ferry's warning that "as a political movement, ecology will not be democratic." Ferry believes that, "as a political movement," "ecology" stands for certain morally preferred policy outcomes. Prioritizing those outcomes, he thinks, implies a willingness to sacrifice "democracy" to their achievement. In gathering evidence for this judgment, Ferry rummaged around the library stacks of English-speaking ecologism. In the process, he picked up on a common feature of "Anglo-Saxon" environmental debate: a proclivity to ponder the relationship between "nature" and "democracy" as a *dilemma*. French liberals could easily take on this leaning because they— like English-speaking ecologists generally but unlike most French ecologists—share the centeredness assumption. Up to a point, then, French liberal critiques of ecologism resemble those familiar to English-speaking readers. But only up to a point. Whereas the shared centeredness assumption makes arguments between English-speaking ecocentrists and anthropocentrists mesh smoothly, in France the liberals' anthropocentrism gets no traction against a local ecocentrism. Liberals veer off into charges of "antihumanism" and ecofascism that largely miss the stakes of the arguments around them. To appreciate how rejecting centering itself might reorient the debate about democracy and ecology, it is helpful first to compare how controversy unfolds in a linguistic community where anthropocentric ecology lines up against its equally centered alternative.

Robert Goodin (1992: 16) contends that green political theory "aims first and foremost to produce good green consequences." Measuring good green consequences requires a measuring stick, a green theory of value. According to Goodin's version of that theory, what makes things valuable is "the fact that they have a history of having been created by natural processes rather than artificial human ones" (ibid., 27). Democracy, in this view, is a theory of agency, not of value. Its worth, from a green viewpoint, depends— contingently—on its ability to produce good green consequences. Should agency and value conflict—for example, if democracy turns out not to protect natural things—it is value that must take priority (ibid.: 120).

Michael Saward (1993: 63–64) worries that Goodin's distinction could justify the use of nondemocratic means to solve ecological problems. Saward's worries are not merely speculative; eco-authoritarians do exist in the English-speaking world. Robert Heilbroner and Garrett Hardin are often mentioned as examples (Press 1994: 12–13; Barry 1999b: 195).

William Ophuls gave eco-authoritarianism what may be its clearest formulation. Ophuls declared that the technical and scientific complexities of ecological issues often put them beyond the ken of average citizens. Placing more power in the hands of an ecologically competent scientific elite is inevitable. "The steady-state society," he concluded, "will not only be ostensibly more authoritarian and less democratic than industrial societies of today . . . , but it may be more oligarchic as well."[3] (Ophuls and Boyan 1992: 215)

Such reasoning is exactly what liberals fear. That is why liberal-minded anthropocentric ecologists search for ways to mitigate the conflict of democracy and ecology. Typically they highlight the vast array of environmental interests that citizens can bring into collective deliberations. Creating an ecological society is not only a technical matter. Aesthetic judgments, considerations of justice, and sensitivities to cultural continuity have to orient the use of techniques to preserve the environment. A democracy in which ordinary citizens articulated such values could be robustly ecological (Sagoff 1988: 166–170). Maybe, but maybe not, ecocentrists reply. Anthropocentric reasoning still leaves species and ecosystems as mere instruments of the popular will, ever vulnerable to the changing whims of democratic majorities.

Anthropocentrists convinced by that critique take the argument to the next level and try to establish a *noncontingent* connection between a more rigorous ecologism and democracy. Thus Michael Saward (1996: 83–88) seeks to prove that, like freedom of speech or the right to vote, a right to health care is a fundamental *democratic* right. Therefore, in demanding a safe, undegraded environment, greens are simply exercising that intrinsically democratic right. Presumably, by arguing for a legally enforceable right to an undegraded environment, Saward would withdraw much of environmental policy from the vagaries of legislative decision making.

But we must watch the nonanthropocentrists most carefully to appreciate the difference between English-speaking and French ecological debate. How can ecocentric greens who favor democracy avoid Goodin's prioritizing of green values over democratic agency? Democracy is rule by the *people*. Doesn't asserting values whose center is *not* the people inherently set up a contradiction with democracy? Not necessarily, asserts Mike Mills (1996). The contradiction may arise more out of consequentialist reasoning than out of nonanthropocentric value theory itself. It is because Goodin and others focus on ecological *results* that their theories of value are in tension with democracy. Mills proposes to replace ecocentrism's usual orientation toward consequences with an orientation toward legitimate political *processes*. Just as democrats gradually expanded the political community to women and people of all races because they came to recognize their moral worth, greens now argue that various holistic, nonhuman entities have moral worth too. Democracy is thus morally compelled to enlarge its participant base. Democracy and ecology can be reconciled, Mills contends, if greens aim to bring those nonhuman entities into the procedures of democratic representation rather than to support only particular policy outcomes. He would give representatives of nonhumans standing in courts, and he would have them sit on regulatory agencies and in the legislature.

In a similar spirit, Robyn Eckersley suggests reconciling ecology and democracy through a reformulated rights discourse. It is widely recognized that individual human autonomy (i.e., the right of individuals to determine their own affairs) and democratic rule are noncontingently connected. This is because nondemocratic government by definition interferes with individuals' judgments of their own affairs, whereas a regime respecting *self*-governance does not. But Eckersley believes that liberal democrats have been self-serving and inconsistent when restricting rights to human beings. The same moral reasoning that leads us to respect human beings would, if applied consistently, also apply to nonhumans. Thus, a corrected liberalism would recognize "the freedom of human and nonhuman beings to unfold in their own ways and live according to their species life" (Eckersley 1996: 377). Liberal freedoms would then have to be realigned in ways "consistent with ecological sustainability and. . . biodiversity." With such reasoning, Eckersley claims, nonhuman life forms would be protected by the very same principle that rules out authoritarianism in human communities: the right of each species to live according to its species life. For humans, this includes the anti-authoritarian "right to choose their own destiny."

Most anthropocentrists and nonanthropocentrists fervently hope that democracy and ecology are mutually supportive. Still, one senses in their sometimes tortuous arguments a lingering unease: a worry that ecological ends and democratic means fit together imperfectly at best. Saward's anthropocentric derivation of green democracy from a right to health is highly contentious from a democratic point of view. An open-ended right to health might absorb enormous quantities of social resources, thereby foreclosing other popular policies. It also fails to get at many environmental issues (e.g., resource depletion and protecting endangered species) that are not health related. As an ecocentrist, Eckersley tries to maintain that nonhumans have a right to their species' life, just as humans have a "democratic" right to self-determination. But the argument is strained. It is incorrect to imply that every human right to self-determination is "democratic." Observing the right to religious liberty, for example, limits the legislative prerogatives of democratic majorities. That is a good thing, but it is not especially democratic. Anyway, as moral limits of this sort increase in number, the scope of popular rule necessarily shrinks. If nonhuman species have a right to self-determination, humans lose their right to govern their own affairs in some areas. There can be no assurance that broadly participatory processes will yield decisions conducive to nature's flourishing. The gap between democracy and ecology ends up having to be filled by some nondemocratic authority.

In preparation for understanding how French liberals treat green theory, one other observation is crucial: *The centeredness assumption underlies every position in this debate.* It is because anthropocentrists and nonanthropocentrists are similarly convinced of the distinct identities of "humans" and "natural" entities that they are so sensitive to the potential for conflict between democracy and ecology. First, both sides in the ecology/democracy debate suppose that nature has a fixed identity. It matters little whether "nature" consists of evolving, interdependent plant and animal life processes unfolding in interaction with physical cycles in the earth's lithosphere, or whether it is seen as "wilderness," or whether it is seen as individual biota that are born, grow, live, and die according to their own manner of being-in-the-world. In each case, "nature" is defined as something with an identity independent of humanity or as something with an objective condition of well-being. That is why democracy is potentially dangerous to it: In pursuing their human goals, members of the *demos* may disrupt nature's identity. In addition, the two sides take a similar view of human subjectivity. As ethicists, theorists treat themselves and their audience as centered individuals. They use their reason to fashion a comprehensive, internally consistent framework of rules to regularize and moralize the domain of human interactions with nature. Centered ethicists tend to think of reason as a capacity enjoyed by each person to evaluate evidence dispassionately, to subject arguments to tests of logical consistency, and to explain events in the world without recourse to unobservable causes. When confronted with incompatible value claims, they try to assess those claims impartially. It is because democracy supposes precisely such an understanding of human subjectivity that green values might endanger it: Nature's good, whatever it might be, seems to trump the values of a being defined by its ability to fashion and live by its own rules.

That, in fact, will be the French liberal ecologists' most important lesson. They make little effort to lessen the tension between ecology and democracy. If anything, they sharpen it. The puzzle is, why do they do so more by wrangling with English-speaking thinkers than with the theorists of French ecologism itself?

Liberal Criticisms: The "Ecofascist" Temptation

In the English-speaking world, deconstructing liberal subjectivity is largely an academic sport, and a rather recently organized one at that. The individualistic and rationalistic assumptions of liberalism are so widely diffused in English-speaking culture that they have rarely had to face mass political movements grounded in alternative theories of subjectivity.

French liberals have had very different experiences (Lilla 1994: 4). Over the last 200 years, they have faced off time and again against tradition-worshipping conservatives, against self-described nihilists, and against fascists claiming inspiration from Nietzsche. Sometimes liberals have been a minority struggling defensively against political movements schooled in

Marxism. Even as France's political order became more and more liberal in practice after World War II, professedly anti-liberal philosophical fashions such as existential Marxism, poststructuralism, and the Situationism of 1968 attracted the allegiance of many intellectuals.

Thus, it is understandable that when some French intellectuals began to speak openly in the name of liberalism in the 1970s and the 1980s, they were combative in their rhetoric and prickly in their defense of democracy. They were quick—too quick, as we shall see—to interpret new challengers as if they were among their familiar adversaries. French ecologism has found itself on the receiving end of those sensitivities.

Among "the new French liberals," as Mark Lilla calls them, is the author of *The New Ecological Order*. Luc Ferry first made his reputation by criticizing "anti-humanist" tendencies in the "thought of sixty-eight." He charged that the seeming individualism of the period of student unrest "led quite predictably in the direction of the disintegration of the Ego as autonomous will, the destruction of the classical idea of the subject." Notions of a personality that is spontaneous and psychically fragmented replaced the "classical idea" of the ego as something that integrates the self and masters its contradictory impulses. This, Ferry contends, has dire repercussions for liberal theory: "The Ego that loses self-mastery, no longer tends to regard other people as other subjects . . . with whom (intersubjective) relations can take the form of a reciprocal recognition of freedoms." (Ferry 1985: 64–65) Since the reciprocal recognition of freedoms is the central philosophical principle of a liberal, democratic society, the "thought of sixty-eight" endangers the core values of the republic. Ferry's subsequent works defend liberal rights and representative government by way of a Kantian "nonmetaphysical humanism."

In a more historical vein, Marcel Gauchet's 1985 book *The Disenchantment of the World* attracted considerable attention in intellectual circles by lacing liberalism with worries about the fragility of moral progress. Modern conceptions of human subjectivity, Gauchet argues, are closely linked to Christian views that God made man free and relieved him of submission to pagan gods. Through long struggles between church and state, this view of freedom has gradually led man to dispense with God altogether. Democratic individualism stands in God's place, for good and for ill. Though democratic individualism is the realization of free subjectivity, it is also dangerously vulnerable to temptations to use state power to do what God was supposed to do: create a perfect society. Gauchet's liberal project is to remind secular moderns that retaining freedom requires abandoning pretensions to godlike wisdom and power.

Ecologism and the new French liberalism developed roughly contemporaneously. But liberals hardly bothered to take notice of green thought until the early 1990s. Before then, greens were generally treated as likable but naive dreamers ("friends of little birds" was the popular, dismissive phrase) who were hardly worth criticizing. Only when Les Verts gathered increasingly respectable scores in elections around 1989 and after the demise of communism did warning sirens begin to sound. The appearance of Serres's book *The Natural Contract* in 1990 raised the stakes still higher. Green politics now had a proponent at the highest level of French intellectual life. Liberals delved into their arsenal. Some outlandishly heavy weapons were deployed.

The philosopher Michel Onfray pounded ecologists by aligning them with the guardians of fascism. "At the ecologists' banquet table," bemoaned Onfray, "you can meet Heidegger and Michel Serres, Philippe Pétain and Antoine Waechter, Ernst Jünger and Félix Guattari." Their ideas all fit together: the critiques of technology, the rejection of Descartes, the love for Rousseau. Ecologists seek an impossible "spotlessness, purity." Like old Pétainistes and former leftists nostalgic for "a time when life was good," greens adopt the myth of "the peasant who regenerates the social body." Like the anti-immigrant National Front party, they aim at a society that is "healthy and vigorous, finally freed of its parasites" (Onfray 1992: 20–21).

An equally scattershot blast came from the geographer Philippe Pelletier. In his view, ecologism, like fascism, combines revolutionary and conservative rhetoric, vitalism, and attachment to the land. Its penchant for sociobiology lapses into racial determinism; its environmental catastrophism prepares the way for centralized, authoritarian government (Pelletier 1993: 88–113). Summing up the combat-weary mood of France's political center, Bernard Oudin (1996: 203–204) sighed "liberal democracy has no luck." It had only just recently routed communism, but "now, on the horizon of the twenty-first century, a new menace springs up: the threat of an ecological totalitarianism."

Assaults by Marcel Gauchet and Luc Ferry employ a slightly lighter, more mobile artillery. In a brief essay that gained attention out of all proportion to the evidence sustaining its argument,[4] Gauchet (1990) updated the liberal critique of radical politics most famously promoted in France by Raymond Aron. For Aron, political radicalism—Marxism especially—is the expression of an essentially religious (read: irrational, millenarian) world view (Anderson 1997: 64–67). Until the fall of communism, it was radical leftists who rejected the world as it was and sought to impose a vision of perfect harmony in its place. Now ecologists, says Marcel Gauchet, are taking their place. Today, with democratic capitalism victorious, Red-green critiques of industrial modernity "recycle theological baggage": they "denounce the satanic pride of a creature that wants to go beyond its limits and mistreats what does not belong to it." Gauchet senses a hatred for humanity in many contemporary expressions of a love for nature. Greens evoke "the edenic dream of a nature freed from the scourge of man . . . , the aspiration of a universe returned to its primordial virginity" (Gauchet 1990: 280–281).

How is it possible for greens to be guilty of recycled progressive leftism and reactionary proto-fascism at the same time? Ferry claims to see through the apparent contradiction. He argues that greens' aspirations to exercise political power could be justified only if they had a philosophy that generated a new perspective not only on environmental questions but also on the full range of issues that governments must address. The only candidate for such a philosophy, Ferry maintains (1992a: 267), is "deep ecology." Only deep ecology, with its notion of the intrinsic value of natural things, has a world view that reorients the evaluation of every issue of governance. What makes deep ecology politically dangerous, in Ferry's eyes, is its tendency either to deny human freedom (by seeing societies as governed by the natural laws of ecosystems that encompass them) or to impugn it (by accusing it of wantonly destroying nature).

In a chapter on "Nazi ecology," Ferry uncovers deep ecology's philosophical ancestry in laws and writings of the Third Reich. Animal-rights laws, a sentimental valorization of primitive nature, and a critique of urbanism and racial assimilationism all have precedents in Nazi Germany. The common thread tying such views together with ultra-leftist egalitarianism, says Ferry, is the hatred of humanist "modernity." Modernity signifies respect for universal human rights and democratic self-determination, appreciation for the power of science and technology to improve the human condition, and acceptance of mobility and change. Ecologists' "obsession with putting an end to humanism" reveals that their "roots lie in Nazism," even as their "branches" extend to a "cultural left" that also misunderstands democracy (Ferry 1992a: 180). Strangely, as his French critics noticed, Ferry mentions only Guattari as a representative of that French left and says nothing of countless green arguments about the nature of work and the quality of life in productivist societies (Saint-Upéry 1992: 146; Deléage 1993a: 13).

Even the much more nuanced views of Dominique Bourg, France's most prolific and ecologically informed liberal environmentalist, tend to combine a reading of English-speaking sources and warnings of authoritarianism (Bourg 1996b).[5] Bourg, Director of the Department of Technology and the Sciences of Man at the Université de Troyes, objects that deep ecologists threaten to "dissolve human dignity in an equal consideration for all things," and thus risk "tearing down the barriers that protect us from ourselves" (1992a: 92; 1993b: 236–237). Deep ecologists infer norms directly from nature and confer rights on nonhumans. Their theory would be dangerous in juridical practice. Nonanthropocentric ecologism would have to take the form of some *people* (the supposed representatives of mute nature) prosecuting other *people* for violations of those rights—a troubling scenario in the eyes of liberals vigilant to prevent unchecked accumulations of power (Bourg 1996b: 50). True, since deep ecologists place the human species as a whole on a moral par with other species, they do not repeat the logic of Nazi racism. But that hardly makes them harmless. In fact, warns Bourg (1996b: 125), deep ecology's hatred for humanity is "potentially much more noxious than Nazism."

Charges of ecofascism in France are too common to be discounted as calculated partisan denigration, but neither should they be taken entirely at face value. Undeniably, French political traditions include a nationalist, racist, ultra-conservative stream (with Maurice Barrès and Charles Maurras as its main theorists) that extols peasant life "rooted" in the fertile soil of *la France profonde*. Such ideas indeed found their way into the ideology of the Nazi collaborationist regime of Marshal Pétain.[6] More pertinent, it is true that there have been occasions on which a few Green Party figures took stands defending regional identities that made them sound intolerant and potentially authoritarian (Paraire 1992: 31; Nick 1991). But none of the main thinkers behind French ecologism has the slightest connection to or sympathy with the traditions of French right-wing thought. The majority of them come from the political left. And it should be remembered that a critique of eco-authoritarianism is *integral* to the works of René Passet, Edgar Morin, André Gorz, and Jean-Paul Deléage, as well as to the writings of all the personalists. Jean-Claude Guillebaud (1992: 20) got it exactly right in his inquiry into French ecological ideas: the ecologists' internationalism and anti-racism, along with their epistemologically nuanced use of ecological science, put them "at the opposite end of the spectrum from rightist naturalism."

What, then, accounts for the frequency of liberal charges of "ecofascism"? Bruno Latour offers an important insight. Moderns, he says, understand passing time "as if it abolished the past behind it." Any reminders of that abolished past become threatening. "If we aren't careful, [moderns] think, we are going to return to the past, we are going to fall back into the dark ages." (Latour 1991: 93–94) For French liberals, the dark age is France's pre-industrial, ruralist past—a past most recently and most distressingly embraced by the Vichy regime. Vichy kept alive a rightist interpretation of a "humanized nature": a nature of tradition and myth, supposed racial "purity," immobility (both social and geographical), relative technological backwardness, a world of "natural" hierarchies in family and politics.

The point is not only that an ultraconservative interpretation of humanized nature is culturally available in France. More important from a philosophical viewpoint, liberals seem bent on reducing alternative ecologisms to that view, *because their own political identity is tied to its rejection.* That is, French liberal humanism gets its meaning *by contrast* to a rightist interpretation of "nature." This happens in two ways.

Ferry clarifies the first of these by tracing his intellectual lineage to the Enlightenment and the French Revolutionaries' Declaration of the Rights of Man and Citizen and explaining that man's *humanitas* "resides in his freedom, in the fact that he is undefined, that his nature is to have no nature but to possess the capacity to distance himself from any code within which one may seek to imprison him" (1992a: 46). Referring to "*any* code" allows

Ferry to assimilate rightist traditionalism and ecologists' ideas of nature's intrinsic value. Substantively dissimilar, both nonetheless devalue human freedom in the same way. To prize custom and social ascription is to deny individuals the capacity to plan their own lives. To recognize the rights of animals, trees, and other things "natural" is to reduce man to merely one right-bearing creature among many. Either way, the special qualities of human autonomy are discounted in a way that, if accepted, disable the liberal project of creating a society of free individuals (ibid.: 173–174).

The second aspect of liberal humanism established by contrast to an ultraconservative nature is a commitment to scientific inquiry and technological progress. Although usually denying that scientific methods can reveal values in nature, liberals have often lauded the role of empirical observation, quantification, and methodical analysis in "disenchanting" the world. Scientific methods have torn down veils of superstition and have rid social practices of futile metaphysical controversies. Additionally, liberals are grateful for the material amenities made possible by science. Not only do these amenities serve human health and satisfy desire; in addition, material abundance helps divert passions that might otherwise feed into socially destructive behavior.

Bourg defends the second aspect of the liberal heritage. Whereas Ferry's environmentalism is grounded in a notion of purely psychic autonomy, Bourg defines humanity in environmentally interactive terms. Paralleling in some ways the analysis of Moscovici's *Human History of Nature*, Bourg (1996a: 9) implies that human interaction with matter is the real foundation of cultural—including moral—development: "Humanity has . . . constructed itself on the outside, on the basis of an edifice that is exosomatic, artificial, and objective, that is, by constituting a network of artifacts that are as much linguistic as utilitarian. Technique, on the one hand, language and knowledge on the other, are both constructions outside of our bodies." By arguing that the ability of the human species to manipulate the environment conditioned the development of its symbolic capacities, Bourg makes ethics itself flow from the qualities of "man-as-artificer" (1998: 117–119; 1996a). Understanding man as artificer rules out any thought of joining deep ecologists in imagining that there ever could be some ideal equilibrium between our species and nature (Bourg 1996b: 104). There must be no a priori limitations on technological development, even as we learn better how to refrain from disrupting ecological processes that are essential to human welfare (Bourg 1997: 71). Liberal nature, in sum, remains a passive substratum which humanity continuously and freely fashions in order to suit its purposes.

As unabashed anthropocentrists, French liberals take positions that seem identical to those of anthropocentrists in the English-speaking world. But the dualism of nature and humanity plays itself out differently in France, especially in comparison to countries with long-standing traditions of wilderness preservation. In North America, for example, calls to "protect nature" can easily be visualized as a geographic distinction; it is a matter of isolating certain nonhuman phenomena (territories and species) from human influence (Van Wyck 1997: 83, 101, 158). In a sense, preserving wilderness is not a project *of* society; it is a project of protecting "nature" *from* society. English-speaking anthropocentrists and ecocentrists often agree that wilderness preservation should have a high priority in an ecological politics, even if they disagree over how much wilderness should be saved and over what sorts of human activity are compatible with maintaining an area as wilderness. But they think of wilderness as nature endangered by the expansion of human ways that are foreign to it. It is much harder to think of nature that way in the thoroughly humanized French landscape. In France, the dualism of nature and humanity takes the form not of a physical separation of two domains but of *politically* distinct societies—different "states of nature," as Moscovici put it. True, liberals understand nature to "[come] spontaneously into existence, independent of any human intervention" (Bourg 1995: 121). But its *significance* always arises in relation to one social-political project or another. Interpreting that significance entails searching for historical precedents, piecing together telling tidbits of political rhetoric, and extrapolating social consequences out of tentative speculations. In France, liberals assume that protecting species and wilderness for their own sake is not really what greens have in mind. In truth, they believe, ecological debate can only be about choosing among various models of a humanized "nature" and among the corresponding types of society—ruralist or "modernist"—that produce them. French liberals associate nonanthropocentric ecologism with fascism because they see both as social projects that make "nature" a pretext for devaluing human freedom.

"Democratic Ecology" in the Liberal Mold

"Democratic ecology" (which is what Ferry and Bourg call their alternative to any project of radical ecologism) unfolds in three movements. It begins with an assertion of the uniqueness of human autonomy. It then attempts to reconcile such autonomy to the severity of the environmental challenges facing humanity. It concludes with appeals to use conventional democratic procedures—with at most a minimal role begrudgingly accorded to green political parties—to address environmental issues incrementally.

A liberal approach to ecology starts from an analysis of human freedom in relationship to "nature." We have seen that liberals fear that ecologism would justify curtailing human rights. The fundamental error of deep ecologists, Ferry charges, is to misunderstand why human beings have a moral significance that generates duties of mutual respect. Lawrence Johnson (1991: 116–118) and some other ecocentrists argue that moral significance is grounded in *interests*. Since nonhuman things have well-being interests, they too have moral significance that demands respect. Ferry disagrees. Moral respect is grounded, he says, in *autonomy*, which nonhumans do not have. Following Kant and Rousseau, he maintains that what makes persons into non-animalistic, moral beings is their ability to use their liberty to check their instincts and egoism. For human beings, autonomy means being able to create new values and to devise impartial rules of conduct. Truly moral decisions are made *disinterestedly* (Ferry 1992: 62).

In a liberal-republican political order, the very notion of citizenship rests on this understanding of morality. Laws—including environmental ones—are morally legitimate because they are connected to autonomous citizens in two ways: they are enacted by a legislature *freely* chosen by citizens, and they reflect the citizens' *general* interests (not merely the immediate appetites of some fraction of the population). No plant, nonhuman animal, or ecosystem can plausibly be said to possess the attributes that confer independent moral significance.

In only one respect might there be a need to rethink representative democracy's humanist foundations. Ferry (1992a: 29, 44–45, 72–73, 126) concedes that *Cartesian* humanism is too concerned with making man "master and possessor of nature" to support an environmentally sensitive liberalism. Cartesian humanists treat everything nonhuman as an unfeeling object suitable for any human use whatsoever. From that perspective, no moral objection to bullfights or to experiments on unanesthetized animals seems credible. Ferry's alternative to Cartesianism is "nonmetaphysical humanism." This would allow us to protect nature, but only for reasons compatible with the uniqueness of humanity's intrinsic value. Many of those reasons, no doubt, will pertain to our interests in our own health, safety, aesthetic satisfaction, and so on. Still, Ferry admits that such considerations may seem insufficient to justify moral concern for some nonhuman entities (e.g., the welfare of animals in general, or perhaps the integrity of relatively remote ecosystems).

The nonmetaphysical humanist then offers an additional line of defense: We can be moved to protect parts of nature when we perceive in nonhuman things qualities analogous to those we respect in human beings. Animals and ecosystems display purposiveness, beauty, even intelligence. These make them "the object of a *certain* respect, a respect which . . . , we *also* pay ourselves," says Ferry (1992a: 124). That we should give animals "a certain respect" does not mean that they have legal entitlements in their own right or that they have intrinsic value. "It is," Ferry claims, "*the ideas evoked* [in humans] by nature that bestow value upon it." Still, recognizing the moral status of ideas evoked through nature's analogues to human autonomy avoids the Cartesian habit of dividing the world into humans (who deserve respect) and nature (which is comprehensively instrumentalized). Ferry's democratic humanism embraces an environmentalism that is "concerned with respecting the diversity of the orders of reality" (ibid.: 123). Presumably, if certain species and natural sites stir our compassion or admiration, legislative protection should follow.

Ferry managed to write an entire book on the dangers of political ecology while barely mentioning the dangers associated with the environmental damage humanity is inflicting on the planet. Bourg may disclose more about the potential of liberal democratic ecology because he is far more attentive to the severity and uniqueness of our ecological challenges. Recognizing limits in the ability of markets to handle ecological problems, he disputes directly the contention of economists that there is (or will be) a substitute for every environmental good (Bourg 1997: 82). Bourg (1996b: 114) sees that resource depletion and climate change raise formidable questions about whether today's democratic citizens are fulfilling their obligations to future generations. He even concedes that democracy, as a form of authority organized in nation-states, is not especially well adapted to tackling global environmental problems. Democratic citizens in one country may well choose to allow pollution as long as air and water currents move the poisons along to their neighbors. Ecological problems such as ozone depletion are all the more difficult for democracies to handle, Bourg notes (1996b: 7–10), because they are invisible. Unlike unemployment or international aggression, they do not automatically trigger people's political concern.

Furthermore, technological developments are transforming the very terms of democratic choice. We have gone from using technology to extend access to familiar goods (e.g., lighting homes through electrification, making agriculture more productive through mechanization and chemical treatment of crops) to using biotechnology to change the very substance of our bodies and the products we consume. Earlier choices serviced age-old desires for material comfort and health but left unchanged other aspects of the human condition (e.g., the ineluctability of death, the randomness of genetic variation). When biotechnologists offer to alter the givens of human identity, the quality of choice changes fundamentally. The limits on what can be desired fall away. The grounds on which democratic citizens will choose to regulate the manipulation of human embryos, for example, are anything but clear (Bourg 1997: 44).

Because Bourg takes notice of unprecedented challenges in the contemporary relations of humanity and nature, his "democratic ecology" is less permissive than Ferry's. Bourg urges that we observe the "principle of precaution" already adopted in a number of democracies (including France): Where a technology or a policy carries significant environmental risks, we should act to prevent them, even if we are uncertain about the likelihood of a particular danger materializing, and especially if anticipated tangible benefits are modest (ibid.: 60, 62). What if the danger is to other species? Alas, says Bourg, technical man cannot make a general commitment to protect every species. Such a commitment would be inconsistent with the technicist vocation of *Homo sapiens*. Nonetheless, concern to safeguard ecosystemic health can justify saving many species, and in addition we can extend our sympathies to those that most resemble humans (ibid.: 108).

Politically, "democratic ecology" amounts to leaving the task of devising environmental policy in the hands of the people's representatives. Not

just any representatives, however. In spite of the French liberals' commitment to pluralistic democracy, they give a decidedly chilly reception to
greens who enter the arena of party competition. The geologist Haroun
Tazieff[7] reasons that a party, in the proper democratic sense, aims for power
in order to implement its program. A truly political program addresses
"everything that concerns the city." But the ecologists' desire to protect the
environment does not touch "everything." Ecologists have no more credibility as a political party than would a "party for employment, a party of
national education . . . , a veterans party" (Tazieff 1989). The political scientist François Goguel too deplores the idea of a "party that has only one
idea in its head [and is thus] unable to contribute to the global political education of citizens" ("Une vision simpliste des problèmes," *Le Figaro*, March
29, 1989). He counsels ecologists to influence public opinion and the government by means of environmental impact studies. Ferry (1992a:
267–268) and Bourg (1997: 125–126) concede that greens serve a useful
social function, but mainly as a "pressure group" expressing a sensibility
especially preoccupied with matters of health, quality of life, and preserving the diversity of beings in the nonhuman world.

The liberals' assumption that "democracy" is synonymous with the
national legislative process and their coolness toward green partisanship
probably have a common source in their view of rational subjectivity.
Although French liberals are not Rousseauean literalists (true Rousseaueans
would favor direct democracy, not representative government), they do
believe that, through democratic debate, the conflicting particular wills of
a diverse citizenry get widened and rectified so that something like a general will of the nation takes shape. Partial associations—metalworkers'
unions, vintners' lobbies, birdwatching clubs—have only a particular will
in relation to the interest of nation as a whole. For this reason, their activities are morally suspect until they have been reconciled to everyone else's
activities through the process of negotiation at the most inclusive level of
association. In practical, political terms, it is the legislative process that
makes autonomous but often self-interested individuals into disinterested,
law-abiding citizens.

Unlike personalists, who distrust political representation as such
(because it inevitably distorts individuality), contemporary liberals regard
it as morally optimal under modern conditions. In large, populous states,

the only practical way to maintain a people's self-governance is by means of a system of representation in which parties aggregate people's preferences across a broad range of policy options. A political party, on this view, should not be just a large interest group. A party is only a step away from directing the legislative agenda of the nation as a whole. To be a fitting candidate for exercising that power, it must offer up a credible outline of the general will in its own program. But greens, say French liberals, are only an environmental interest group (unless they are unconfessed eco-authoritarians). Hence, they would be unable to moralize autonomous citizens by drawing their wills into a disinterested program.

In essence, liberal democrats in France insist that ecological politics be conducted under three fundamental philosophical stipulations. First, there must be no a priori moral judgments about the intrinsic value of natural things, for such judgments might underwrite a jurisprudence that would override democratic decisions. Human rights come first. Second, policies should be formulated without anti-scientific prejudice against technology and "artificialization." The very definition of human freedom includes its ability to alter its surroundings. Third, "democratic ecology" must be interpreted as a matter of negotiating how environmental issues should be more consistently integrated into the processes of representative democracy.

Liberals believe that people have an abundance of prudential reasons to want to protect nature. Thus, when they summon the will to push environmental measures through the legislative process they practice "democratic ecology." Of course, today's citizens also desire inexpensive consumer goods, imported commodities, individualized transportation, secure employment, and the right to build a house in an unspoiled country setting. Some political parties will, no doubt, attempt to come to power by aggregating those competing interests. If they succeed and then decrease the funding of the Ministry of the Environment or loosen the definition of wetlands to facilitate property development, that will be "democratic ecology," too. Such is democracy.

Unanswered Questions of Liberal Environmental Concern

What does "democratic ecology" really stand for, ecologically speaking? Have liberals in any way advanced debate beyond reaffirming their faith in representative government, whether its policies are "ecological" or not? The unfortunate consequence of French liberals' preoccupation with fending off ecocentric authoritarianism is that it has kept them from engaging the major issues raised by a noncentered ecologism. Let us consider a philosophical shortcoming and then a more practical one.

Referring to issues such as preserving wild species and protecting the interests of future human generations, Alain Lipietz (1993a: 16) responds to Luc Ferry with a reminder that "political ecology raises problems settled by no social contract, no fundamental pact between free individuals." Why do such problems go unresolved? Let us reexamine Ferry's reasoning. He regards man as "the antinatural being par excellence" because human autonomy consists in the potentially moral capacity to resist "natural" impulses. In this Kantian worldview, egoism—the rule of appetites and emotions—is our "natural" lot (Champetier 1993: 76). But as "autonomous" beings, we establish our *own* purposes. We do not take them from nature. Moreover, what is "respectable" in other human beings is not their cultural identities or their physical attributes; it is their autonomy. Autonomy is moralized when people agree to act according to general principles. Mutual recognition means that one autonomous subject does not impose ends on others. Respect forbids coercion.

The unique challenges posed by ecology do not translate easily into this type of discourse. Certainly when Ferry calls us to "respect the diversity of orders of the real" the meaning of "respect" has changed in a mysterious way. The nonhuman things we are to respect—say, blue whales and salt marshes—do not order their existences according to values they have *chosen and universalized*. They cannot *reciprocate* respect. Thus, even if they are autonomous and purposive in *some* way, they are not autonomous in the *particular* way that demands respect according to Kantian theory. Nor do Ferry's noninstrumental reasons for valuing natural things offer a new and compelling manner of conceiving respect for nonhumans. To protect a species because we find it beautiful or playful is to make its survival depend on the vagaries of our culturally shaped emotional response to it. We will try to save the pandas, but we will probably be indifferent to the survival of less lovable (or tastier) species. How can anything called respect be so arbitrary? The sense in which we should "respect" nature remains entirely unexplained.

Bourg's sensitivity to the novelty of issues created by modern technologies brings him closer to the threshold of a noncentered ecologism. Cartesian-Baconian science, he says, divided reality into a nature of non-teleological, causal relations, and a purposive humanity. As a result, "man and nature [could] no longer be conceived together" (Bourg 1997: 18). Today, however, we are witnessing a "reinscription" of man in nature and of nature in human society. Intellectual frameworks such as Darwin's theory of evolution, cybernetics, and systems ecology contribute to "naturalizing" human behavior, while we find ourselves taking charge of more and more processes that formerly were spontaneous ("natural")—as when we make decisions to protect the biosphere or modify genetic materials (Bourg 1995: 116–126). As clearsightedly as any of the other French ecologists, Bourg perceives that conceptions of "nature" have changed in ways that force us to rethink what it means to be human. "We belong to nature," he stipulates, "in the sense that we come from it and cannot possibly live outside of the biosphere" (ibid.: 126).

Then Bourg adds this: "We nonetheless occupy an eminently unique place, relatively exterior to that same nature, a place that confers upon us our technicist status." (ibid.) When the liberal arrives at the threshold separating centered from noncentered humanism, he refuses to cross. For liberals, nature remains "relatively exterior" because subjects remain centered relative to a discoverable external world. They reject nonanthropocentric ecologism out of fear that it would undermine the premises of democracy and technical progress. They reaffirm the status of the centered self and a distinct nature even when admitting that our acting according to our "technicist status" carries a genuine risk of destroying a humanly habitable biosphere (Bourg 1998: 82, 116–117; 1995: 136).

A second set of unanswered questions concerns the liberal position on property in an age of environmental limits. Not only ecosocialists but also most other French ecologists, ranging from the systems theorist René Passet to the personalist Philippe Saint Marc, would agree with Alain Lipietz (1993a: 42) that "the main adversary of political ecology is . . . economic liberalism." French ecologists identify liberalism with a doctrine that justifies unlimited appropriation of nature in order to satisfy human desires, stimulates resource-depleting consumption, disregards the intrinsic diversity of things by giving them all a price, creates incentives to transfer environmental costs and risks onto those who did not create them, makes labor tedious and powerless in the name of "efficiency," and insists on international free trade even if it undermines domestic environmental legislation and exacerbates economic inequalities between nations. Surely one of the most important questions we might pose to liberal adversaries of French ecologism is whether their "democratic ecology" must end up tolerating such effects in the name of preserving individual liberty. In France only a tiny, minority school offers a clear answer: free-market economists of a strain that Americans would call libertarian.

"The majority system," complains the economist Gérard Bramoullé (1993b: 49), "favors the right to vote over the right of property." Bramoullé finds that favoritism completely "abnormal." Libertarians hold that individuals express their uncoerced preferences in market transactions (Bramoullé 1993a: 35). Government regulations—even those backed by a clear majority—interfere with free exchange. Restrictions on building, species-protection acts, and waste-disposal regulations trump property owners' decisions and threaten violators with coercion. Libertarians need not deny the existence of environmental problems. They decry situations where one owner pollutes another's property. They acknowledge that it is undesirable for people collectively to exploit a resource out of existence. Such problems, argues Max Falque (1986), prove only that a system allowing such things to occur does not really rest on private property and the free market.

In the libertarian perspective, Garrett Hardin's 1968 essay "The Tragedy of the Commons" demonstrated that collective goods problems arise because property rights have been left unspecified (Falque 1986: 42; Bramoullé 1993b: 38; cf. Barry 1999b: 150). People overexploit resources when goods are freely accessible to all rather than available only at a cost to legally specified property owners. The benefits each person receives individually exceed the share of costs he or she must bear from overuse. The existence of external diseconomies is solved by fixing responsibility—by specifying property rights. When downstream fisherman have a right to clean water, for example, upstream polluters will have to either reduce their pollution or compensate the fishermen. Where property rights are clear, the parties will have the financial incentive to seek the most efficient allocation of resources.

Government regulation is presumptively wrong, because it is tantamount to coercion. Instead of allowing agents freely to settle their conflicting interests in the ways most suitable to their particular situations, the state threatens them with penalties to force their adherence to general rules. Unnecessary recourse to regulatory action, moreover, engenders inefficiency and abuse. Pressure groups influence government policy to favor their partial interests. Bureaucrats, who are self-interested actors like everyone else, end up more concerned with extending their power than with effectively administering programs (Falque 1986: 48, 52; Bramoullé 1993b: 42). Because collective-goods problems are real, *some* government regulation is inevitable, libertarians concede. But a democracy that respects property rights will minimize regulation by seeking market-based solutions whenever possible.

Libertarian reasoning displaces the main issue. The ecosocialist Alain Lipietz tracks it down. To Bramoullé's habit of contrasting "property rights" *tout court* to the coercive power of the state, Lipietz responds that the libertarian position really just sets aside the political issue of defining the scope of property rights. Taking account of that definitional issue reveals that state regulation is not the opposite of individual freedom and indeed is often its precondition.[8] Lipietz distinguishes three types of environmental problems that call for regulation beyond that spontaneously generated by markets (Lipietz 1993b: 40–43).

Some of these problems stem from what Lipietz calls "composition effects." These are situations where noxious results arise only from the simultaneous (composite) actions of many people. The use of chlorofluorocarbons by a few individuals does not endanger the ozone layer; their use by many people does. With composition effects, libertarians' paradigmatic example of injustice—one person's harming another by violating his or her property rights—breaks down. Where composition effects are possible, an individual's action may or may not constitute "harm," depending on whether others engage in the same action. Property owners can hardly be expected to manage composition effects, since as individuals they are not clearly responsible for them. Regulatory intervention is required.

Second, private property transactions are limited in their capacity to take account of long-term consequences. Capitalists seek profits in the short term. In a competitive market, pesticide manufacturers cannot afford to wait to see if their new products affect animal life in 20 years. Monoculture yields higher returns on agricultural investment than organic farming, even if it risks destabilizing local ecosystems in the long run. Ecological processes usually take place on a much longer time scale than market processes. Where we have no adequate substitute for goods provided by ecological cycles, only a nonmarket authority can safeguard them.

Problems of the third kind concern collective goods. Now, libertarian theory admits that genuinely public goods do exist. Some state intervention to protect public health and essential resources is therefore unavoidable (Falque 1986: 53). But collective goods are more elusive than that response appreciates. What exactly constitutes a "problem" is a matter of great contention and variability. Suppose one person has property rights in a piece of land. Lipietz (1993b: 40) asks whether ownership gives that person the right to construct buildings of any height or to broadcast loud noises. Adjacent landowners may claim that enjoyment of pleasing vistas or neighborhood silence are rights inhering in *their* property. Whether they make those claims depends on their tastes, their sensibilities, their feelings of entitlement, and their understandings of health and risk. But if they do assert such claims, no principled conception of liberty or property settles the question. The need to regulate them, Lipietz concludes (1993b: 45), "justifies in certain cases, [the intervention of] democratically controlled public authorities."

Though France's new liberals might quibble with some aspects of Lipietz's regulationist ecologism, they do not seem inclined to disagree with him in principle. They appear to accept the Rousseauean—and very unlibertarian—premise that democratically determined communal values take precedence over the prerogatives of property. Luc Ferry (1992a: 237–238) imagines a "democratic ecology" generating interventionist legislation while barely pausing to consider the libertarian fear that majorities interested in saving species or landscapes will trample on the rights of property owners. Claude Allègre, a professor of geoscience and the author of several books explaining the science of ecological problems to a general public,[9] is skeptical of environmental catastrophism. Yet Allègre perceives serious problems ahead for the planet: a lack of consumable water, widespread pollution of the oceans, air pollution in cities. With a candor that might astonish readers in countries where politicians such as Ronald Reagan and Margaret Thatcher have made liberals loath to defend the state too vigorously, Allègre (1990: 373) declares that "the state will play a capital role [in dealing with environmental problems]. As legislator, it will have to set rules of interdiction (as few as possible!) or tolerable norms. As economic regulator, it will have to . . . set the price of the environment and monitor fair practices in the private sector."

French liberals find little attraction in libertarian theories that try to deduce democracy from individual property rights and which therefore limit the state's power in relation to property. Many take it for granted, as Rousseau did, that the state embodies a set of general interests before which self-interest must give way (Duclos 1989: 140–141). In this regard, their differences with ecosocialists are a matter not of principle but a matter of degree. Yet they give no clue to what that degree might be. As a result, liberals' cries of eco-authoritarianism lose much of their punch. Dominique Bourg blames environmental degradation on an "economic ideology" of individual property rights—rights exercised independent of nature's capacities for self-renewal. He declares unhesitatingly that the state's "classic role" of protecting citizens now must extend to "fundamentally new domains and unavoidably contradict certain liberties [previously] arranged for economic agents" (Bourg 1998: 97; cf. 1994–95: 134).

Perhaps even "liberals" will one day countenance a level of environmental regulation that today's property owners would denounce as oppressive. Or will they? They may decide that economic freedom is too precious to allow extensive regulation. "The ecologists never stop demanding more numerous forms of public intervention, more draconian regulations, more severe penalties," complains Bernard Oudin (1996: 205). Indeed, ecologists have problematized issues that potentially touch every aspect of life—the intellectual as well as the practical, the public as well as the personal. Environmental controversy now surrounds land-use patterns, waste disposal, scientific research, decisions on human birth and death that affect population levels, genetic modification of foodstuffs, every form of resource extraction, and the treatment of animals. Does liberal freedom immunize individuals from collective norms in all those areas? In any of them? On the other hand, are liberals who dislike regulation willing to defend the consequences of laissez faire policies? Or is the only way to salvage their defense of autonomy to deny the seriousness of environmental problems?[10] Until they address the issue of the relation of freedom (especially economic freedom) and nature more directly, French liberals have barely begun a debate with their indigenous ecologism.

Ideas for a Noncentered Democratic Ecologism

The liberals' assumption that democracy is synonymous with the formal legal practices of contemporary Western governments further restricts their ability to take up debate with French ecologists. In the past, "democracy" often was a protean "protest ideal" that encompassed far more than well-established procedures for connecting public policy and the popular will (Sartori 1987: 337). Activists have sometimes urged the "democratization" not just of the state but also of factories, schools, and even the home. Democracy has also been held to describe a way of life—wherever instituted—characterized by an aspiration to make political and social equality more effective, by a conviction that assessments of the quality of life are too personal and too precious to be left to an elite, by a commitment to the value of deliberation, and by a devotion to norms of tolerance and social inclusiveness.

Ever since the student rebellion of 1968, French ecologists have drawn creatively on those traditions. Serge Moscovici (1978: 127–129) expressed the spirit of French ecology's early enthusiasts when he praised an "ecological nebula" of citizens who would form small groups to raise public awareness of pollution, of the dangers of nuclear power, and of urban crises. Alain Lipietz (1993a: 28–30) proposes a "participatory ecological democracy" based on face-to-face negotiations between organized groups of environmentally affected citizens. Jean-Paul Deléage (1992a: 9–10) would bridge the gap between technocratic environmental decision making and the understanding of ordinary citizens by encouraging the formation of a culture of "counter expertise." A physicist by training, Deléage (1993b: 10–12) nonetheless refuses to concede to ecological science a "presupposed objectivity, completely external to society." He would make visible the disagreements and inevitable political-moral assumptions of scientists by organizing the confrontation of different groups, each with their own "partial objectivities" (ibid.). If there are dangers of "ecofascism" in these ideas, one never hears French liberals make the case.

Claims that ecology could foster a politics that is more participatory, more anti-hierarchical, and less adversarial than what is ordinarily practiced in liberal democracies are a common feature of green discourse around the world (Goodin 1992: 139–146). In France, an additional type of claim sometimes takes shape. In the absence of a polarized debate between two ecological centrisms, demands to "democratize nature" turn into projects of resituating citizens relative to the processes through which "natural" identities and "humanity" get constituted together. Bruno Latour and Denis Duclos illustrate two ways of imagining such a project.

Latour faces up to Ferry's charge that ecologism is undemocratic. The liberal, Latour observes (1993: 25–26), detects among greens a tendency toward intellectual closure that would "be done with the indeterminacy that is essential to democratic questioning." Latour responds that if Ferry wants artifice and uncertainty he would be hard put to do better than to extend the ideas of Serres's natural contract. *Politicizing* nature in the manner of Serres and Latour requires envisioning some means of joining the "mediating" practices of science with the arbitrating practices of democratic governance.

Recall that Latour does not regard the ethic of the natural contract as an imperative deduced from certain timeless, morally relevant properties of "natural" things. Rather, he sees it as a new stage in the production of hybrids (1994a: 796). Humans now apply lessons learned from managing large-scale technologies to a world plagued with disorders. If we accord rights to nonhumans, it is because contemporary hybridization has made it possible to conceive of nonhumans as having subject-like properties. They act creatively; they play complex roles; they seek stability. Latour's "parliament of things" assembles all those who can give voice to those properties and makes them negotiate with representatives of various human interests. Its function is to "manage an ongoing experiment in which the capacities, degrees of resistance and properties of the ensemble of human beings, collective persons, rational beings and nonhumans are tested through a series of trials." (Latour 1994b: 110) "The parliament," Latour continues, "looks much more like a laboratory than a congress, but this laboratory looks much more like a forum or a stock market than a temple of truth." The parliament of things is a sort of institutional hybrid. It combines and revises elements of politics, science, and administration. From politics the parliament takes the model of representation. Open-ended debate, reconciliation through negotiation, and uncertain outcomes are integral parts of this model. Scientists now speak for nonhumans within a larger assembly. They maintain commitments to systematic research, but as co-negotiators they cannot simply announce Truths about the world to other representatives. From the practice of administration the parliament of things takes the function of maintaining records, ensuring consistency of procedures, and conducting evaluations.

The idea of a parliament of things makes us think about democracy and "nature" in ways that depart from the strategies of centered green theories. Anthropocentrists usually do not problematize the construction of "nature" itself. Michael Saward's argument for a democratic right to an undegraded environment, for instance, takes scientists at their word when they assert that a given alteration of "nature" does or does not endanger human health. Whether to allow the use of cattle feed that includes processed sheep offal (to take a recent example) depends on the state of scientific research on the safety of such practices. Absent contrary evidence, beef from cattle fed processed sheep by-products is deemed safe for human consumption. Later, when bovine spongiform encephalopathy ("mad cow disease") spreads and gets linked to a degenerative brain disease in humans, our democratic right to health (and health is a manifestation of our "nature") mandates a change in policy. Along the way, "nature" has clearly changed. Yet anthropocentric ecologism simply accepts current wisdom about its properties.

Latour's parliamentary proposal is an attempt to get us to abandon the assumption that scientists unproblematically define "nature" for us and that regulatory authorities objectively handle scientific data in order to protect the public welfare. In effect, when sheep by-products were fed to cattle, a massive, open-air experiment was launched upon the public without ever being a matter of political discussion. Latour's ecologism, in contrast, stimulates us to investigate connections among scientists, regulators, and commercial interests that affect how "nature" is made to appear at any given time (Latour, Schwartz, and Charvolin 1991: 52). Latour wishes to expose those connections to the light of public examination. He would oblige decision makers not only to answer for their errors after the fact but also to answer for their expectations at every point in devising the experiment which they propose to run: when planning to introduce new techniques, when measuring their effects, when statistically evaluating data, when enlarging the geographic scope of the experiment, when extending its duration, and so on.

Problematizing the construction of "nature" in this way also makes Latour's parliament differ from Mike Mills's idea of placing representatives of nonhumans in licensing authorities and regulatory agencies. As an eco-centrist, Mills is committed to distributing representation as a function of an entity's end-orientation. Latour, in contrast, endeavors to design a form of representation whose purpose is to make visible and debatable the very process through which entities become identified *as such.*

Latour suggests one way in which democratizing nature might involve far more than raising the prominence of ecological concerns in the legislative chambers of democracies, as liberals propose. "Hybridization" issues have heretofore been handled in relatively closed settings by small numbers of people with a limited range of interests. Imagining a parliament of things entails inventing new ways of eliciting and transmitting people's expecta-tions of "nature" as they form at every site where decisions jeopardizing those expectations are made. Latour's ecologism has a claim on the adjective 'democratic' because he would make the production of hybrids something that is "regulated and decided in common" (1991: 194). He taps into the intuition that, in a democracy, what affects all should be decided by all.

Denis Duclos, too, tries to envision a "democracy" whose citizens are not unified subjects squaring off against "nature." Recall that, for Duclos, each of us is psychically divided in our relationship to objects and different con-ceptions of "nature" arise in our attempts to give expression to these divi-sions. No unifying rationality can overcome these divisions. In fact, it is various forms of rationality—economic, scientific, moral—that threaten to suppress individuality when they are pursued obsessively. Civility gives each of our various object-relations its due, granting supremacy to none. Duclos depends on an equilibrium of passions, not on the reign of reason, to civi-lize human relations—and to protect "nature." Civility leaves space for a "nature" that is untamed, unrationalized, independent, unique. Duclos (1996: 169) argues that political subjects need such a "nature" to exist phys-ically around them in order to shore up their own sense of freedom.

Duclos argues for a "democracy of the passions" in order to secure space for people's self-expression in the face of powers that are reshaping every nook and cranny of the world in their image. He worries especially that global networks of economic exchange are drawing every region and every aspect of daily life into the maximizing logic of capitalist production (1996:

108). Democracy, for Duclos, is not simply a form of government that institutionalizes majority rule. Even less is it to be confused with the supposed diffusion of power in the marketplace. Democracy is a form of government that respects the undetermined individual. Democrats recognize that the substance of political life consists of what individuals bring into the political sphere *from the outside*—their singular "natures" as beings with inscrutably different, utterly incommensurable sources of pleasure [*jouissance*]. It betrays democracy, Duclos charges, when such singularity is denied effective expression—as when liberal policy analysts place all potential users of the countryside on the same plane, all with comparable interests (1996: 213).

To respect singularity more fully, it would be necessary to translate civility into a new spatio-temporal distribution of human activities. Duclos imagines a utopia in which each of our principal passions would have a territorial correlate. Some space should be given over to utterly wild, unmanaged nature; there would be linguistic regions in which individuals formed their cultural identities; self-governing cities would constitute places in which citizens worked out a limited sharing of goods and functions while respecting one another's private delights. Meanwhile, carefully contained channels of universal exchange (of commodities, information, energy) would connect the various centers of population (ibid.: 243–260).

Duclos (ibid.: 237, 308) invites us to think of this territorial division as analogous to Montesquieu's idea of the separation of powers. Montesquieu (another thinker schooled in skeptical humanism) had little interest in asserting an abstract concept of freedom. Freedom is simply what people do when left to pursue their own inclinations. His famous doctrine of the separation of powers supposes that a space for such freedom is best secured by balancing the branches of government against one another. Each branch has a degree of autonomy from the others and operates according to norms that constitute its special role in the scheme of governance (e.g., in the case of the judiciary, making judgments in an impartial, rule-bound way). Similarly, Duclos's utopia is designed to create different staging points, as it were, for our nature-intimating passions, gathering the force of each of them in a concrete way so that each can effectively assert its "singularity" against the intrusions of the other. Rather than a world of "generalized connection" driven by commerce and technology, Duclos (ibid.: 276) envisions a situated life: a world of sites, each having "a relative autonomy resulting from the organic relation of diverse functions, concretized in a place." This "republic of sentiments" would constitute a realization of democracy. In existing liberal democracies, "merchants of political goods, without faith or law" dominate political life. A regime of territorialized passions, in contrast, rests on "cities that are controllable by the ensemble of citizens" who are dedicated to "linking the duration of their life to a particular place" (ibid.: 277).

To avoid misunderstanding Duclos's theory, one must constantly remind oneself that "nature" always appears to us through symbolic grids that we have devised in order to cope with passions directing us toward unshareable objects. What justifies the territorial divisions of a republic of sentiments is not a concern to protect wilderness "for its own sake" or even a concern to set aside ecosystems that are vital to Gaia's health. Those sorts of motives presuppose that "nature" has an objective existence apart from our passions. Duclos's point is that we cannot get at "nature" directly. We can do so only indirectly, by practicing various forms of concrete disengagement—various types of "retreat from domination." The territorial divisions of a republic of sentiments give substance to such withdrawals.

As was discussed earlier, Duclos criticizes Latour for failing to articulate the need for any limits to hybrid production. In a sense, Duclos faults Latour's relentlessly experimental pragmatogony for bypassing anything like the "principle of precaution" that even liberals (ecologically sensitive ones, at least) endorse. Yet the liberal version of precaution fails to impress Duclos, too. He suspects that, as a principle, precaution will seldom prevail in a world where nations are bent on constantly advancing commercial interests and spurring scientific inquiry. Anyway, applying such a principle seems to presuppose the existence of constituted agencies capable of evaluating contradictory technical evidence about whether a proposed technology might be dangerous. Decisions about nature are once again shunted into the hands of experts and government authorities. The principle of precaution, in other words, partakes of one logic: the logic of the scientifically controlled technical system (Duclos 1996: 148).[11] To avoid an environmentalism of "geo-management," we must redefine harm. In a reinterpreted principle of precaution, "harm," now no longer conceived as this or that modification of an ecosystem, would consist in interference with the various "natural" conditions for the expression of irreducibly distinct forms of human subjectivity—conditions which it is the purpose of a territorial republic of sentiments to secure (ibid.: 266).

If one approaches the ideas of Duclos and Latour by way of a detour through English-speaking ecologism, it is the similarity of their intellectual strategies that stands out, not the two thinkers' disagreements. Neither man could be mistaken for an ecocentrist. Neither thinks of ecological theorizing as a rule-creating enterprise, a sort of anticipation of the legislative process. Both speak approvingly of a new ecological humanism without buying into anthropocentrism. Most important, neither is searching for an abstract value or vantage point that might reconcile competing claims about the ultimate "nature" of environmental concern. Rather, both propose to *reposition* citizens relative to trends that simultaneously constitute them as subjects and "nature" as an object of concern. This repositioning—in Latour's case, getting citizens to participate in a parliament of things; in Duclos's, attaching them to territories that represent different facets of their divided psyches—is not designed to pre-decide the priorities of democratic decision making. It seeks to put citizens in a better position from which to see and to judge processes (such as the "mad proliferation of hybrids" and the invasion of scientific and economic rationality into every domain of life) that are actually generating today's widespread unease about the survival of "nature."

The Liberal Challenge

This is not to suggest that Latour or Duclos have cut the Gordian knot of green theory. Difficult questions are coming. However, for the moment a simple point is in order. It is that these noncentered theories, questionable or not, have barely attracted the attention of French liberals. English-speaking liberals, in contrast, have risen to the green challenge. Some maintain that an expanded notion of a "commons" protected by public power has considerable purchase on environmental problems (Wells and Lynch 2000: 86–87). Ronald Paehlke (1989: 191–192) argues that, by granting basic rights to future human generations and to other species, "environmentalism can be taken as extension of liberalism." Such calls for an *extended* liberalism take seriously green claims that traditional formulations of an ideology designed to protect individual liberty against the encroachments of power are not sufficient to address ecological problems. English-speaking liberals concede that it is not easy to account for moral intuitions about the wrongness of eradicating nonhuman species or disrupting complex ecosystems if one makes the enhancement of human autonomy the standard of political right. Not easy, but perhaps not impossible either. Apart from some of Bourg's writings, advocates of French liberalism have kept the conundrums of ecological politics at arm's length. They have yet to generate anything comparable to Marcel Wissenburg's (1998) carefully reasoned liberal defense of "the free and the green society." Their principal contribution to ecological debate lies in their having clarified just how crucial the humanity/nature distinction is to a liberal interpretation of humanism. Descartes's dream that we would one day become the "master and possessor" of all nature was not a parenthetical outburst of enthusiasm from a man struck by science's potential to conquer the ancient scourges of humanity. It was the *logically necessary complement* to a particular conception of personhood—one that emphasizes our status as free, thinking beings capable of taking a universalistic perspective. If Ferry is right, that conception of personhood is fundamental to the moral framework of a liberal democratic society. It underwrites the case for including all people who are endowed with such freedom—not just some "naturally" gifted elite—in the polity's legislative processes. It prescribes limits on how power may rightfully be exercised. Respect for freedom proscribes measures that create arbitrary distinctions among citizens. Thus, it is no accident that the liberals John Locke and Immanuel Kant invoke the idea of a "state of nature" on the way to justifying a *civilizing* politics of representative government, individual rights, and the rule of law. A democracy that can legislate across the whole range of issues that become matters of human concern (including environmental ones) while remaining sensitive to matters of personal freedom and technological innovation seems to presuppose a notion of reason that is *inherently* in tension with nature.

If this is the case, then greens who demand that we reconceive reason to make it more friendly toward nature face a stiffer challenge than they sometimes acknowledge. Nonanthropocentrists often say that they want reason to be more holistic than analytical (Naess and Rothenberg 1989: 33, 63), more accepting of our continuity with other species (Baxter 1999: 232), more willing to embrace the idea that moral significance attaches to things other than the human capacity for self-control (Johnson 1991: 43–96). At the same time, they assume that this revised reason can coexist with democracy, with respect for human rights to life, with free expression, and with self-governance. At least, green arguments explicitly rejecting such rights are rare.

Rationalistic French liberals dare greens to demonstrate the consistency of such a combination of ideas. Once one argues, for instance, that well-being interests rather than rationality establish the moral significance of living entities, the groundwork is laid, not for democracy, but for paternalistic governance by elites who can best discern the interests of their mute charges. Once one adopts the holistic principle that value is diffused throughout an ecosystem, the notion that rights inhere in each individual begins to crumble. The danger of ecocentrism, from the French liberals' perspective, lies as much in its internal contradictions as in its overt prescriptions. Unimagined sacrifices loom in some green polities.

In spite of their insight into such philosophical dangers, however, French liberals remain frustrating. By concentrating their criticisms on "Anglo-Saxon" variants of green theory, they fail to perform what would be a signal service in their own linguistic community. If they would redirect their fire toward domestic greens, they might force them to explain how a case for democracy can survive an attack on views of a rationally discoverable nature and on a rationally autonomous humanity—views that have been used to support democracy since the seventeenth century.

In fact, noncentered theories go even further than centered ones in challenging conceptions of reason. After all, Latour and Duclos do not argue that reason can be extended or amended. They doubt that it is capable at any point of integrating human experience into a self-consistent, morally comprehensible whole. Latour adopts an epistemology in which disorder, flux, and random variation are irreducible characteristics of human experience. Reason does not *find* sense in this chaos. At best, it *makes* sense. It weaves partial, disconnected webs of meaning out of arbitrarily spun strands of concepts. Duclos too emphasizes the "arbitrary" character of civility (1993a: 246; 1996: 215). Civility skips among the symbolic registers of identity, convention, and science. *Why* it skips must remain unexplained and enigmatic; otherwise civility itself would be a social *logic* with its own potential to become a self-enclosed *puissance*. A belief that ecologism, at some level, requires transcending *logic* itself is what makes Duclos, as well as Latour and Serres, rely on an open-ended process of discussion or of balancing passions to achieve ecological effects. And these ecologists are hardly alone in their thoughts about the limits of reason. From Morin's view of *Homo sapiens demens* to Charbonneau's theory of a psyche that is always in tension with the world, the picture of a divided, noncentered self seems to rule out a notion of reason that could master disorders.

Thus, the French liberals are right to suspect that French ecologists challenge liberal suppositions about reason, even though they are wrong to believe that greens do so by asserting "natural" values. "Contingent" networks of scientific inquiry and "arbitrary" passions are the real sources of the French ecologists' challenges. Still, it is not hard to imagine what liberals would say if they confronted noncentered ecologism more directly. They would probably charge that theories of the ultimately arbitrary organization of knowledge are just as inimical to democracy as green theories of value.[12] Almost in spite of themselves, the liberals manage to pick out a problem that noncentered theorists address inadequately. Defending humanism and democracy seems to require an appeal to reason. After all, doesn't any debate about our environmental future require evaluating evidence, criticizing logic, and seeking out general principles of rightful conduct? If so, one is using reason—the center of our being, say liberals. If not, it seems that any future is as good as any other, regardless of its effects on either people or nature and regardless of whether it is democratic or not.

How French ecologists might respond to such a dilemma is the topic of the final chapter.

8

Questioning Nature: Reason and Skepticism in French Ecologism

One possible way to reconcile green theory and democracy is to show that it is *liberal* democracy, not democracy per se, that copes inadequately with ecological problems. Alain Lipietz, we have seen, chides *liberal* democracies for so anxiously defending private property that they cripple the community's ability to impose effective measures of environmental protection. *Liberal* democracies, says Val Plumwood, tolerate economic inequalities that allow the privileged and powerful to buy their way out of many environmental ills; meanwhile, "the most oppressed and dispossessed people in a society are . . . made to share the same expendable condition as nature" (Plumwood 1995: 138–139). *Liberal* conceptions of the public/private distinction obscure how new political issues, including environmental ones, take shape as social groups contest the exercise of power in what before was seen as the private sphere (Hayward 1994: 204).

An increasing number of green theorists advocate *deliberative* democracy as an ecologically rational alternative to liberal democracy (Brulle 2000). Often inspired by Jürgen Habermas's discourse ethics, deliberative democrats criticize liberals for reducing democracy to a process of aggregating people's existing preferences (Dryzek 1987: 201; Barry 1999b: 217). They offer reasons and evidence to think that deliberative democracy would set in a motion a process of preference formation in which green values would demonstrate their superiority (Gunderson 1995: 10; Torgerson 1999: 120; Dryzek 1994: 192–194). The democratic ideal, they say, should be collective action sustained by a dialogue that is as free as possible from partialities and distortions (Hayward 1994: 205). Most of them also agree that linking ecologism to Habermas's communicative ethics unavoidably makes green theory anthropocentric.

If a reconciliation of democracy and ecology through Habermasian discourse ethics rests on the centeredness assumption, one must expect non-centered French greens to resist this extension of environmental thought—as several of them do. But if one gets beyond the rhetoric of centeredness, it is possible to see considerable (if still imperfect) overlap between many forms of French ecologism and insights generated in a communicative action perspective. I intend to show how some of the claims that French ecologists try to express but leave undeveloped can be elucidated with a Habermasian account of ecological rationality. What French theorists would gain from a Habermasian interpretation is an account of the conditions under which their claims about an ecologically more rational society can have the public validity that they assume them to have.

On the other hand, Habermasian ecologism—once interpreted independent of the rhetoric of centering—could take on a significant insight drawn from the French: the case for discursive democracy trades on multiple concepts of nature. These concepts resist complete rationalization. Even a discourse ethics, it turns out, depends on divided natures. Recognizing this, France's skeptical humanists supplement the case for a green deliberative democracy. The rhetorical repertoire of the skeptical humanist is what permits deliberating citizens to move *among various* forms of argument that protect the diversities of nature.

Centers Revisited

French ecologists divide into a number of strains of what I have called "noncentered theory." But noncenteredness is not a simple or uniform concept. It admits of degrees and different interpretations—all of which, nonetheless, stand in contrast to the anthropocentric and nonanthropocentric green theories encountered in the English-speaking world. A review of this general contrast and of the various forms of noncenteredness clears the way for an encounter between the ecologism of Alain Lipietz and Habermasian theory. I will use this encounter as an initial test case for determining whether a theory of communicative action can make sense of certain claims that are asserted, but not thoroughly defended, in the works of some noncentered French ecologists. Ultimately I want to show that such a rapprochement is possible because Habermasian discourse ethics itself is not incontrovertibly centered. Since that demonstration depends on a firm grasp of the distinction between centered and noncentered ecologism, I turn first to exploring its meaning and implications.

English-speaking green theorists debate constantly about *where* to locate the center of environmental value. Rarely do they probe the notion of having a center itself. What is the significance of identifying a "center" in a political-ethical context? The first point to note is that the language of "centering" is metaphorical. It transforms ideas of uniqueness and moral measurement into a spatial image. Serge Moscovici (1974: 221) recalls the metaphor's original referents: the language of centering borrows from historians' descriptions of debates in early modern science between the earth-centered astronomy of Ptolemy and the heliocentric theory of Copernicus. These debates were not only factual. One reason Copernican astronomy was so controversial was that Christians had come to see Earth's central position among the planets and stars as an expression of God's intention to give humanity a special place in His creation. To decenter Earth was to demote humanity and call into question prevailing understandings of the source of moral order. A center is not merely a point in space. It is a unique thing endowed with certain properties allowing it to illuminate the value of other things. The goodness of those things becomes a function of their proximity to the "center."

Cartesianism clearly puts humanity at the center of a moral system, leaving the analytically disaggregated components of the nonhuman world to take on value in relation to human interests. Orbiting close to the center, a natural entity—say, a flounder—is brightly lit. But species of no direct interest—say, plankton or kelp—are like distant and dimly lit celestial bodies. Almost beyond the limits of sensory detection, they pass darkly through the furthest reaches of our moral space.

An ecologically enlightened anthropocentrist corrects certain features of the metaphor while preserving all its internal relations. Humanity remains the sole source of light in the moral system, and natural things continue their revolutions at greater and lesser distances from the center. But ecological science allows us to perceive complex life-sustaining interactions among what once were seen as separable components of the environment. What gets brightly illuminated as "nature" is now not just the flounder but the whole ecosystem of which it is a part—plankton and kelp included. In addition, enlightenment makes the center burn brighter. Nonutilitarian interests in nature become more evident, and we foresee the need to protect future generations. The strengthened rays of our interests touch the remote wilderness and the glaciers of the Arctic shelf.

In one sense—in relation to such a moral cosmology—nonanthropocentric ecologists rightly claim to "decenter" humanity. Like Copernicus, they show that what had been presumed to be a unique "center" in fact shares qualities with other entities heretofore regarded as peripheral and inferior. Nonanthropocentrists locate value-conferring properties in non-human entities: the fact of being alive, of feeling pleasure and pain, of subsisting as an entity that has well-being interests, of exhibiting "wonderful" qualities such as beauty, intricacy, and elegance (Baxter 1999: 67–70). Such arguments decenter humanity because they mean that henceforth we do not uniquely radiate moral value.

Yet such decentering does not eliminate centers. It really means *recentering*. Moscovici (1974: 225) explains:

Any Copernican revolution that places the "sun"—whatever it may be—in the place of the "earth" unfolds in the same manner: it reverses the terms of a known relation . . . ; it establishes knowledge or the real around the new center . . . ; it causes a theoretical or empirical break, exchanging the actors and modifying the position of the light source that illuminates the stage, whose dimensions and props remain unchanged—as does the play that takes place there.

Many supposedly radical theoretical innovations are really just centers switching places.

That is clearly the case in certain formulations of nonanthropocentrism. James Lovelock (1995: viii) recounts how, as his appreciation of "Gaian" processes of self-regulation grew, he shed his "loyalty to the humanist Christian belief in the good of mankind as the only thing the mattered." Now he sees humanity as "no more than a part of a community of living things" and argues that "we humans have no special rights, only obligations to the community of Gaia" (ibid.). Lovelock has simply reversed the terms of a known relation. Nature is now the sun, and humanity is now one of its many orbiting planets.

Other nonanthropocentrists will object that this image miscasts their meaning. "Nature," they might say, is not a homogeneous entity that other bodies circle. It is a vastly inclusive system of interrelations and interdependencies. The point of discovering value in nonhuman things is to emphasize that those entities shine forth moral "light" in their own right. Their significance does not depend just on illumination they receive from humanity. Perhaps the better image is that humanity is just one star in a firmament aglow with cross-lit astral bodies.

But there are two reasons to believe that a spectral center still haunts such theorizing. Its presence appears, first, in the tendency of nonanthropocentric theories to end up giving special privileges to humanity relative to the rest of nature. (See chapter 2.) Suspiciously, one star in the firmament burns much more brightly than the rest. More important, though, is the fact that nonanthropocentric theorizing really amounts to a search for new ways of fulfilling the same *ethical function* as the old human "sun."

The function of a center is to give things a transitive ordering in a scheme of values. Cartesian anthropocentrism gives us reason to believe that we humans have automatic and uncontestable pre-eminence over all of nature and can decide the fate of natural things strictly as a function of our purposes. Nonanthropocentrists reject that moral scale, but they hasten to put a new one in its place. They accept the assumption that the main task of green theory is to offer ideas about how to manage tradeoffs between human desires to consume and reshape natural things and the moral standing of those things themselves. Whatever moral scale the nonanthropocentrist proffers is designed to situate humanity in relation to nonhuman things with which our values might conflict, and to offer a priori reasons for resolving the conflict in a certain way. A nonanthropocentric scale of moral considerability *is the new center*. Yes, many things, humanity among them, now shine with moral light. They do so, however, for the same reason that objects outside Plato's cave dazzle the enlightened prisoner: because they reflect the illumination of the Good.

A center, then, is a metaphor for moral relations that conveys three fundamental ideas:

• There is a need for a comprehensive ethical system to apportion value among all entities within its reach.

• Those entities have determinate identities. Copernicus and the medieval astronomers did not disagree about what the sun and the Earth were. What is at issue between rival centrisms is the relationship of known bodies, not the very identity of those bodies or their conceptual interdependence.

• As in a circle, there can be only one center. That is why anthropocentric and nonanthropocentric greens engage each other's arguments with such single-mindedness. Both sides realize that one center establishes itself at the expense of the other.

It is a very different type of argument when French theorists examine the temporal succession of humanity-nature couplings mediated by labor and technology, or when they assert that scientific and moral understandings of nature infiltrate each other's images and rhetorical practices. When theoretical assertions take this form, attention gets directed elsewhere than to constructing ever more elaborate versions of environmental ethics. Theoretical endeavors focus instead on uncovering the history, politics, and psychology of the practices that manifest themselves today as the source of environmental problems.

When I label French ecologists "noncentered," I mean to highlight an unusual shift in perspective common to their theorizing. The ordinary shift in green theorizing is from one center to another: Ecocentrists try to convert anthropocentrists to their views, and vice versa. The unusual shift attempted by French ecologists involves staking out a third position enabling them to look at assertions about humanity and nature, as it were, *from the outside*. From the inside, participants in green debates see their competing arguments as efforts to locate the "real" center of environmental concern. From the outside, *those very debates* are seen as *products* of a larger, moving system of relations. To step "outside" is to try to grasp a more comprehensive *process* that drives ecological (and anti-ecological) debate itself.

Such a step, of course, does not really make theorists external to every claim in contemporary debates. They are not at the end of history; their perspective is not God's. Their move is more like a displacement, a relative resituating of vantage points. Still, this displacement is sufficient to make their controversies fall into a different pattern. In the rhetorical field of French ecologism, discussion focuses on whether rival theories have adequately grasped the *process* linking nature and humanity. All sides in the debate agree on one thing, however: Studying this process implies that there is no true center at all.

This unusual shift in perspective is carried out in various ways, with varying degrees of epistemological radicalism.

In its least audacious form, noncentered ecologism is a species of ethical pluralism. René Passet and Philippe Saint Marc claim that "nature" abounds with meanings that no economic measure can translate. Indeed, they deny the very possibility of commensurating values in ways that might allow the development of a formula for regulating relations between human communities and the various "natures" they care about. Pluralists have an external perspective of sorts in regard to ostensibly competing green arguments, for in their eyes it makes no sense to debate over the true center of environmental concern. Pluralists understand that what the participants in those debates experience as contradictory arguments are in fact assertions of irreducibly distinct values, each deserving some place in our moral order.

Those I have associated with "politicizing nature," as well as the ecosocialists and Moscovici, occupy an outside vantage point in a stronger sense. For all of them, there are *historical* processes that link social practices and understandings of nature. Ecosocialists focus on the social evolution of the means of production and accord labor a major role in generating changing views of "nature" and "humanity." These theorists look into the processes of scientific research in order to question their independence from domains of social power and myth making. In each case, the thinkers set out a theory that explains not what nature (or its value) *is* but rather how our conceptions of it and our self-understandings co-develop through time.

One additional external perspective is explored in the works of Morin, Duclos, and other personalists. They perceive processes in the human psyche that call forth correlative "natures." For them, how we think about nature is inseparable from our own drives for possession and recognition. Their lesson for greens is that they must rise above the quarrels of competing centers in order to avoid simply being carried along by the psychodynamics of arguments over power. They must see their own logic as well as that of their adversaries as incomplete and prejudicial in interpreting nature.

Noncentered ecologisms still must face a tough series of questions about their intellectual coherence: *Why* should we adopt their "outsider's" perspectives? In political debate, people usually ask for *reasons* to adopt one course of action rather than another. Can a noncentered theorist—one who relativizes reasons to historical circumstances or who traces them to the contradictory desires of a divided psyche—even offer reasons? The writings of the French ecologists seem to indicate that they can. None of them simply gives up on logic or empirical evidence in presenting his case. But even if their discourse has its reasons, what is supposed to make them persuasive to persons who are differently situated in social, historical or psychological terms? And what sense can be made of the French ecologists' demand for more "democracy"? Is it really possible to justify democratic practices without reliance on the nature/reason dichotomy that has traditionally grounded them?

Discourse Ethics and Regulation Theory

A closer examination of how Alain Lipietz does (and in some cases, does not) answer these questions exposes how his path converges with that of Jürgen Habermas's communicative ethics. Habermas works out in a more complete sense a variety of claims that are integral to Lipietz's regulationist ecologism. A demonstration of this convergence has significance beyond the bounds of Lipietz's theory. Serge Moscovici, Edgar Morin, and other French ecosocialists follow substantially similar routes.

Recall that Lipietz's regulationist ecologism pictures a green program as the precursor of new compromise designed to control the socially and environmentally destabilizing tendencies of liberal productivism. In a regulationist model, such a stabilizing agreement gets forged only through the struggles of diverse groups to impress their needs on reluctant political and economic elites. In chapter 6, I also noted, however, that Lipietz's emphasis on the contingency of compromise, his unwillingness to invoke any notion of historical progress, and the difficulty he has explaining how ecological demands arise according to a regulationist model leave the ethical basis of his green contract mysterious.

Much of the mystery would disappear if regulation theory could be plausibly interpreted as a concrete application of Habermasian communicative ethics. Communicative ethics distinguishes between compromise understood as "a balance of power" and compromise as an agreement incorporating "norms [that] express generalizable interests" (Habermas 1975: 111). So does Lipietz's theory. Interpreted in Habermasian terms, Lipietz's proposed compromise could be seen to issue from a respect for individuals as autonomous, mutually communicating moral agents, capable of evaluating their own circumstances and of negotiating social arrangements that embody a rational consensus. From the viewpoint of discursive ethics, individuals meeting in search of compromise are not bundles of power pitted against one another; they are communicative actors seeking to provide universal justification for their claims about the good of their community. Merely by engaging in this dialogue, they implicitly accept one another as equals, as agents who can understand and act according to general moral rules.

Thus, Lipietz's commitments (and those of other contractual ecosocialists, including Gorz and Deléage) to equality and individual autonomy could be seen as necessary ethical presuppositions of the bargaining situation. The concept of an "ideal speech situation"—one in which participants reach their conclusions only through "a rational redemption of justified claims"—furnishes a standard for disqualifying "compromises" that were results of coercion or ideological manipulation. Communicative ethics would also fit well with the Third Worldism of Lipietz and René Dumont, since its conception of generalizable interests requires us to seek ecological policies that respect inhabitants of all parts of the world equally. Not only does the Habermasian interpretation affirm the same ethical standards that contractualist ecosocialism promotes; it does so by extracting those standards from the very processes that regulationism makes central to its analyses of social change: the processes of negotiation and compromise. Regulation theory as communicative ethics could explain why the idea of a "social compromise" *deserves* to be at the center of social theory.

Habermas makes sense of an even bolder claim that Lipietz advances about the prospects for diverse new social movements converging toward "a single will for change." Seeking an agent of social change and yet rejecting both traditional Marxist hopes for a universal class and a simple conglomeration of protest groups, Lipietz (1988a: 95) concludes that "the only solution is thus to work toward the maturation of a 'shared meaning,' an alternative culture . . . , so that *each can recognize in it not only their own direct interests, but equally the interest that each has for others to find their interest there too.*" Evidently Lipietz aims at a social order suffused with "mature" feelings of reciprocity and community, not merely a grudging willingness to trade off advantages. The ethical basis of regulationism is the extension of interest—interest enlarged and enriched through the development of interactive competence.

Explaining how such a maturation of shared meaning is possible is the theme of *The Theory of Communicative Action* (Habermas 1984). Habermas sees history as a learning process in which humankind accumulates knowledge about the conditions of its fullest emancipation—conditions that require an ever-more-complete sharing of meaning. Only recently have (some) societies (very imperfectly) learned to quell conflict by recognizing the humanity of workers, tolerating ethnic diversity, equalizing economic opportunities, and democratizing political structures. These values are not merely elements of the most recent social compromise—elements that are morally incommensurable with those of preceding compromises. These are *advances* in humankind's ability to abide by norms of reciprocal accountability. History, in other words, reveals progress in our moral consciousness as well as in the area of technical-instrumental knowledge.

This approach is not only normative. As social theory, it offers an important hypothesis about why certain social structures are more stable than others, and hence how social change can take a particular, progressive direction. Modernity, according to Habermas, embodies a "rationalizing" project. As societies try to enhance their capacity for material reproduction, they also find their diverse members interpreting their needs within that system and subjecting them to discursive testing. Along with technological sophistication, societies develop their collective identities in ways that express higher degrees of communicative competence. Discursive testing subjects the society's practices to critical questioning. Are its distributive principles capable of being "communicatively shared"? Are they based on generalizable human interests? Do they encourage critical reflection among citizens? Failure to measure up to such testing motivates system instability. Dissatisfied social actors search to reshape the social system according to more rationally defensible norms. The Habermasian view makes us see that what constitutes "crisis" is not merely a dysfunctioning system. It is a *delegitimized* system. More than that: delegitimization occurs because the system fails to meet the *evolving* norms of social actors who are becoming increasingly competent at criticizing myths and justifications serving partial interests.

There is a significant obstacle to reading regulationism in this way: Lipietz has at times appeared to reject this interpretation categorically. He has specifically repudiated any view of history as a movement through a staged series of social transformations that spontaneously generates an ethically preferred resolution of a community's contradictions (Lipietz 1987: 22). Teleology in this sense, he maintains, is only an illusion. Reflecting on the disappointed hopes raised by Marx, Lipietz (1993a: 116–117) concludes that "progressivism needs to be reinvented." He continues: "It can no longer count on the movement of history, on the development of technique and knowledge, it can no longer be satisfied with praising modernity. . . . Henceforth, progressivism must always be on the side of the poor in the name of an ethic of solidarity . . . , it must take a stand in favor of an *other* modernity. . . ." Lipietz knows that regulation theory makes it appear that the pieces of a model of development fit together as if they were "made for" one another. But this is only because it is an "a posteriori functionalism" (Lipietz 1988b: 35). In fact, no particular compromise is historically preferred; "notions of 'reform' and revolution' are thus relativized" (Lipietz 1988a: 83). Compromise is simply whatever accommodations various groups have settled on in their search to mitigate conflict.

And yet, those interpretations simply fail to translate many of Lipietz's theoretical claims. His own words imply that he *does* believe in some notion of ethical progress. He urges Europe to take the lead in putting forth "better compromises" between economic activity and environmental preservation, just as it earlier led the way to "better compromises between capital and labor" (Lipietz 1992b: 124). In moving from one mode of regulation to another, ethical advance is possible. Moreover, when Lipietz speaks of "better compromises" he parallels a Habermasian argument that moral imperatives for change *grow out of* our experience with the irrationalities of existing institutional arrangements. He argues that the Fordist compromise has been in crisis since the late 1960s *because* (among other reasons) workers whose jobs are precarious and whose workplace-acquired technical expertise is systematically ignored become less productive. He maintains that its productivist commitment to unlimited economic growth generates so much waste and pollution that people eventually come to demand more environmental protection. "At the end of the eighties," Lipietz writes (1989: 153), "the rising social, macroeconomic, and ecological perils are belatedly provoking a new awareness," which supports "more advanced social compromises." In Habermasian terms, one would say that new social movements, sparked by the dysfunctions of prevailing exclusionary or ecologically unsustainable social practices, challenge defenders of those practices to meet the test of the generalizability of their interests.

Before reaching ecological conclusions, however, this analysis must leap one more hurdle. Although a historical learning process might explain how human communities, through a refined ability to take the perspective of the other, acquire a universalistic moral sensibility with respect to interpersonal relations, it is still difficult to understand in what sense we might include "nature" in that sensibility.

There is some evidence that, as Lipietz has become more taken with ecology, he has begun to perceive the problem. In 1989, ecological concerns take the stage in the fifth chapter of his book analyzing the decline of Fordism and the impasse of liberal productivism. Environmental problems appear not as an essential motive for the crisis of Fordism but as an additional constraint on any solution to the problems of rising unemployment and increasing difficulties in financing the welfare state. The crisis of Fordism is first and foremost a crisis of declining production. Without environmental awareness, the solution would seem to be stimulating production and distributing its benefits more widely. New-found awareness of the growing burdens on environmental systems, though, makes Lipietz reject the productivist strategy. He insists that the capital's quid pro quo for workers should be not more purchasing power but more free time. Remaining within the regulationist perspective means envisioning a new compromise— a sort of social contract—assuring all interested parties that their interests have been taken into account in their society's principal distributive decisions. Four years later, Lipietz (1993a: 16) fired this criticism at the *liberal* social-contract tradition: "Political ecology raises problems that no social contract . . . can solve. Thou shalt not kill—whom? . . . 'Your partners in the social contract,' respond secular thinkers. Fine. But then, what about wild animal species? . . . And future generations of human beings? 'After me, the deluge' responds the individualist who founds his ethics on 'self-interest rightly understood.' No expressed interest: no social contract." The same words have a dangerous potential to ricochet and wound regulationist ecology.

Before it was extended to ecology, there was nothing theoretically incoherent about viewing social systems as modes of development stabilized by multiple negotiated arrangements linking workers, employers, and the state. Workers who cannot tolerate the pace of the production line resort to industrial action to force management to slow it down. Managers who believe that workers' wages endanger profitability attempt to get workers to accept lower wages. Systemic balance results from achieving a "compromise" among all the interested parties.

That pattern of argument does not apply so well to the issues of ecological politics. Animal species, pristine territories, and future generations can no more be present in social bargaining sessions than they are in liberal fantasies of a hypothetical contract. They cannot sit at a bargaining table or destabilize institutions by marching in the streets. How can we conceive ecological goals in regulationist terms when the potentially disadvantaged "parties" lack the capacity to represent their own interests?

Lipietz (1993a: 18) responds by demanding a "dynamic extension of an altruistic consciousness." He appeals for social movements to place the interests of future generations, the Third World, and nature into the ethical deliberations of democratic societies. But an "extension of an altruistic consciousness" is not so easily incorporated in a regulationist framework. After all, in the early twentieth century a social reformer might have asked capitalists to "extend their altruism" and identify more with the needs of their workers by giving them shorter hours and health insurance. Regulation theory places little faith in such remedies. The Fordist compromise was hammered out through strikes, protests, and hard-fought electoral battles. Workers pushed through recognition of their needs in spite of the profit-maximizing ethic of their employers. Regulation theory's emphasis on the conflictual nature of group relations leaves little room for good will as a source of social stability. If Lipietz covers ecological concerns primarily by appealing to altruism, it is hard to see that he is still in the regulationist framework.

A Habermasian framework, on the other hand, could readily explain such "altruism."[1] It is a result of the "expansion of the domain of consensual interaction" (Habermas 1979: 120). Steven White sees grounds within Habermas's recent works for connecting a historicized communicative ethics and ecology. Surely it is more than an accident of group struggle that more and more people have developed interests in the protection and restoration of the environment. White (1988: 137–138) proposes seeing "growing environmental crises . . . as a practical catalyst for reflection on how the ways in which we currently assault nature are leading to a more and more frustrating and self-destructive form of life." It is plausible to argue that they have developed these interests *because* the widespread destruction of nature has activated a sense of loss or aesthetic distress or fear of dangers to health. The advance of technical-instrumental knowledge puts nature at ever greater risk, but it also creates the conditions for our becoming more aware of the severity and implications of that risk. Alarm about the global effects of technology and demands for preservation of wildernesses and protection of species now have entered into political deliberations once almost monopolized by the quest to maximize production or to distribute social goods fairly. White (ibid.) notes that those who share these concerns then begin "experimenting with alternative forms of life and technology [with a] potential for enhancement of a sense of balance or harmony with natural systems." Thus, critical reflection and aesthetic yearnings could prompt an expanded "sense of what makes for human satisfaction and well-being." What Lipietz calls "altruism" is really a human interest in the nonhuman world that is becoming more articulated through collective deliberations in which the criteria of communicative competence come into play.

Ecologism is more than just another stage in a meaningless series of humanity-nature relations. Lipietz's ecologism constitutes a movement toward a greater degree of *consciousness* about the conditions of our existence. The growth of consciousness is not something we witness with indifference or treat as a merely contingent development. Consciousness is superior to unconsciousness. This is true intrinsically (what is the point of philosophizing at all if it does not somehow enhance awareness?) and because it is a condition of our coming to grips with challenges whose dimensions we now better understand. There is a conception of *rational* development at work here.

It is this conception of ecological consciousness that reveals additional connections between communicative ethics and additional French ecologists. Edgar Morin and Serge Moscovici, too, foreshadow—but do not really elaborate on—a notion of ecological rationality. Both criticize the Cartesian project of "mastering" nature. Then, instead of renouncing mastery entirely, they propose changing its target. Moscovici says that his "political technology" aims at "the government of the natural order." He calls on humanity to "master a movement, by transforming relations of which we ourselves are a part." For Morin, the goal of political ecology should be the "control of control." What can "governance" or "control" mean in the context of theories holding that nature is not something "out there" that human beings manipulate at will?

Both Morin and Moscovici see movement in our conceptions of nature, in our understandings of human "nature," and in our understandings of external "nature." This movement has a direction: Humanity has only recently gone from taking nature for granted—seeing it as something quite distinct from human subjectivity—to beginning to perceive it as something generated in a process linking an external world, human practices, and ideas. Perceiving this movement allows us to shift our understanding of mastery. Cartesian mastery meant gaining control over nature's forces and making them useful to man (Chalanset 1997: 18). Now, if that very mastery creates dangers of its own, then "mastering a movement" must mean studying the *processes* producing Cartesian mastery and bringing *them* under some more inclusive form of guidance. Since it is through processes of scientific experimentation, technological innovation, and dividing labor that we produce each state of nature, "the government of the natural order" means reflectively undertaking action with respect to *those* processes so that they do not produce dangerous and unwanted effects. Like Lipietz, Morin and Moscovici evoke something resembling Habermas's theory of social evolution as a learning process.

Green Deliberations: Democracy without a Center?

An ecocentrist is now sure to cry foul on the ground that I have defended the intelligibility of a noncentered ecologism by smuggling a center back into the analysis. Habermas's green sympathizers (indeed, Habermas himself) agree that communicative ethics is anthropocentric (Hayward 1994: 208; Whitebook 1996: 284; Habermas 1982: 247). As an ecocentric critic puts it: Habermas's theory "privileges human emancipation vis-à vis the emancipation of nonhuman nature" (Eckersley 1992: 110). There is a straightforward reason for this charge: In a communicative ethics, intersubjective norms arise through unconstrained deliberations of human subjects about their interests. Under those circumstances, people's various environmental sensitivities might well legitimize a wider range of ecologically rational practices than are found in liberal democratic societies. However, Habermas "is unable to work the interests of nonhumans into his work in any meaningful way because it is *theoretically grounded* in human speech acts" (ibid.: 111; cf. Eckersley 1999). Ultimately, human interests alone dictate how nature gets treated.

There is an additional problem facing any search for affinities between Habermas's theory of consensus-oriented interaction and noncentered ecologism. Habermas foresees growing unity in the human species as a collective subject. Though the logic of instrumental rationality obtains in our relations to things, among our own kind we seek to vindicate our communicative interest—our interest in mutual understanding through undistorted communication. This interest, too, has a history: Modernity's progressive "disenchantment" of the world, its tearing away of veils of magic, myth, and prejudice gradually increases the critical learning capacity of the human species as a whole. We become more unified, more universal—more *centered.*

In other words, Habermas posits a teleology. A teleology generates a moral framework that allocates value to persons and nonpersons. In its ultimate stage, humanity becomes conscious of itself as the radiant source of value. In addition, the orientation of the process toward the end determines the significance of each of its earlier stages. Thus, even in earlier stages, when the center to be is still dispersed, the center is already present as the measure of whether the incomplete aggregative process is taking its proper course.

Thus, for better or for worse, it seems that an ecological communicative ethics must be anthropocentric through and through. A dilemma looms for my interpretation: Either French ecologism is noncentered (in which case it cannot sustain a Habermasian interpretation) or it is centered (in which case my protestations to the contrary are in vain).

Yet the source of this dilemma may be less the content of Habermas's theory than the tendency to surrender to the rhetoric of "centering" without first reflecting on what centering means. In fact there are on Habermas's side abundant materials for a rapprochement with noncentered ecologism. Just as I argued earlier that Lipietz makes implicitly Habermasian claims, I now maintain that many of the claims that qualify the French as non-centered show up in Habermas's theory too.

It is important to recognize the modesty of Habermas's teleology. As Thomas McCarthy explains (1978: 239), Habermas's "theory of social evolution . . . requires neither unilinearity, nor necessity, nor continuity, nor irreversibility in history." A teleology that is so relaxed should cause no alarm in a theorist such as Alain Lipietz. It does not deny human freedom. It does not confer dangerous authority upon a prescient political vanguard. It *does* "relativize" notions of reform and revolution, as Lipietz prefers. In contrast to ecosocialists who treat "capitalism" as if it had only one form, Lipietz—like Habermas—insists on recognizing "the existence of a variety of capitalist development models," some of which respond better than others to social inequality and environmental damage (Lipietz 2000: 78). Habermasian evolution certainly does imply "a conception of cumulative processes in which a direction can be perceived" (McCarthy 1978: 239). But this is no different from Lipietz's assertion (2000: 72) that "political ecology defines progress only as a *tendency*—defined in terms of certain ethical or aesthetic values (solidarity, independence, responsibility, democracy, harmony)."

To be convinced of the modesty (and the relative noncenteredness) of Habermasian teleology, one need only compare it to the theory of "dialectical naturalism" that underlies Murray Bookchin's social ecology. Social ecology, according to Bookchin (1989: 37), "establishes a basis for a meaningful understanding of humanity and society's place in natural evolution" by building on a "conception of nature as the cumulative history of more differentiated levels of material organization . . . and of increasing subjectivity." Human critical reflexivity is the culmination of nature's becoming self-conscious. Ecological awareness, for Bookchin, has arrived when humanity builds into its own social practices the characteristics of nature itself: equality, not hierarchy; differentiation, not uniformity; cooperation, not competition. An egalitarian ecological society would, like ecosystems, be self-regulating through decentralized processes of dynamic interaction (Bookchin 1982: 345–353). Habermas's teleology is humble indeed next to this vision. For present purposes, the essential difference is that Bookchin postulates a telos that constitutes a *particular* model for human communities to follow, whereas Habermas hypothesizes only a cumulative learning process that renders human communities increasingly capable of rationally evaluating *any* model of social organization put before them. Bookchin (1995: 132) claims to be noncentered. In fact, his ethics depends on a selective and highly debatable view of what nature *is*. *Real* nature, according to Bookchin, is characterized not by competition or random variation but by symbiosis, mutuality, and equilibrium. In centrist fashion, Bookchin makes "natural" properties cast their evaluative light on human practices. Habermas, although he envisions a teleology of human communication, minimizes his substantive description of a rational society. A rational society, he argues, remains open to discursively testing any proposal to protect nature—no matter what "nature" might be at issue. At the very least, this aligns Habermas's theory with the interpretation of noncentered theory as ethical pluralism, à la Passet or Saint Marc. Moreover, Habermasian greens argue that in a deliberative democracy debatable norms could even include putatively ecocentric ones (Torgerson 1999: 162–63; Brulle 2000: 46). If so, then deliberative democrats—unlike Ferry's liberal democrats—need not check the entry passes of arguments before allowing them to go through the doors of the deliberative chamber. In this respect, Habermasian ecologism would be just like the undeniably noncentered ecologism of Bruno Latour, who looks on with equanimity as today's green activists talk of "granting to nonhumans some sort of rights and even legal standing" (Latour 1999: 202). Not that Latour's *own* philosophy of science definitively validates the understandings of "nature" embodied in such talk. The point is that noncentered theorists are committed to giving such talk a hearing because of its role in a *process* of change.

The process perspective draws Habermas even closer to noncentered theory. Studying a social learning process requires altering one's perspective just as noncentered theorists do. Furthermore, Habermas's description of the modernization process has close affinities with the ideas of France's internationally best-known political ecologist: André Gorz.

Habermas's theory of modernization starts from the lifeworld of individuals—the pre-theoretical world of everyday life, with its linguistically mediated accretion of traditions, narratives, and social roles. The practical resources of the lifeworld are vast, though unsystematic. Modernization is a process in which those resources are, to a certain extent, rationalized. The world is "disenchanted"; religious and mythic orientations toward action are marginalized. Social order differentiates itself into three distinct cultural value spheres—science and technology, morality and law, art and literature—each with its own "specialized form of argumentation" (Habermas 1984: 163–165, 236–240; 1988: 95). Scientists invoke truth in their studies of nature; citizens invoke normative legitimacy in consensus-oriented debate; individuals invoke authenticity in discussions of aesthetic phenomena. This differentiation of argumentative modes marks a decisive improvement in the learning capacity of humanity as whole. Scientific, political, and aesthetic institutions become better able to adapt to new circumstances as their different types of validity claims get separated from one another.

But, Habermas thinks, Western modernization has been "one-sided." Institutions from one of the three available spheres of discourse have become predominant. Under capitalism, market relations and the state administrative institutions corresponding to them have expanded to the point where democratic and aesthetic institutions are less and less significant in steering social evolution (Habermas 1987b: 186–187). Market and administrative relations have now "colonized" so many sectors of life, and have subjected so many decisions to the impersonal logic of economic efficiency and bureaucratic categorization, that they endanger the very independence of the lifeworld. The subjugation of the lifeworld to a restricted set of steering mechanisms risks the obliteration of modernity's capacity for critical renewal and adaptation. In relation to ideas encountered in the French context, two things about this analysis are noteworthy. First, a Habermasian analysis of our environmental predicament and André Gorz's analysis have a strong mutual resemblance (Little 1996: 24; cf. Bowring 1995). Gorz sees political ecology as a response to extensions of economic and administrative rationality that come at the expense of the lifeworld (1993); his contractual ecosocialism rests on the hope that ecological issues can be addressed in a more fully rational sense if we can bring the cultural resources of the lifeworld to bear on them through deliberative political processes. Since Gorz is an acknowledged influence on Dumont, Lipietz, Deléage, Duclos, and many other French ecologists (Roose and Van Parijs 1991: 81–82; Jacob 1999: 297), and since Habermas has become a reference point for more and more English-speaking ecologists (not to mention German ones), Gorz's common cause with Habermas's explanation of environmental awareness builds a sturdy bridge that might allow greens to cross more easily between linguistic communities.

The second point to emphasize is that Habermas's critique of the colonization of the lifeworld is noncentered in the most important sense explained earlier. It obliges the theorist to step "outside" the debates as contemporary participants frame them. For example, ecological activists may protest drilling for oil in the Arctic National Wildlife Refuge. They may frame their critique in terms of the sacredness of wilderness or the folly of risking irreversible damage to an ecologically fragile area merely to get a short-term supply of gasoline for American commuters. Those are centered claims—claims apparently grounded in nature's intrinsic value or in the interests of future (human) generations. But the theorist does not simply side with one or another of these positions. He sees both of them, from the "outside," as reactions to the "one-sided" character of modernization. His *own* position then gets defined only in relation to an understanding of the larger process at work. If a Habermasian ecologism seeks to reinforce institutions in which dialogue is not skewed by money and power, if its analysis of discursive designs favors protecting "autonomous public spheres" (Dryzek 1994: 189; Torgerson 1999: 10; Hayward 1994: 205), it is because of a conviction that such measures will facilitate the uptake of people's reactions to environmental stresses and put them to work in devising more rational approaches to ecological issues. It was from just such a perspective that Moscovici (1977: 555) defined the task of an ecologically oriented "political technology" as to "allow people both to direct their collective destiny and, by foreseeing their own evolution, to prompt the emergence of its successive stages."

Divided Natures in a Green Deliberative Democracy

However, noncentered ecologists advance some arguments that have no parallel in Habermas's theory. By *reciprocally* problematizing nature and humanity, noncentered ecologists confound Habermas's schema of modernization. Nonetheless, French ideas, drawn from the tradition of skeptical humanism, could help compensate for deficiencies in Habermas's conception of nature.

Noncentered ecologists repeatedly transgress the boundaries of what Habermas regards as modernity's distinct complexes of rationality. In the view of Michel Serres, scientific validity claims already contain particularized and politically contentious suppositions. (For example, Serres distinguishes between a violence-inclined science that uses "martial" images of power and competition and a Lucretian physics of flows, freedom and relations.) Moreover, claims about political legitimacy cannot be separated, either conceptually or practically, from claims about the structure of the world. (For example, Serres ruminates on how "contract" works its way into both scientific understandings of nature and political conceptions of reciprocal obligation.) If knowledge structures are as crossfertilizing as Serres contends, validity claims too will reach across cultural spheres.

Latour (1991: 81; 1999b: 318–19) charges that Habermas widens "the abyss between objects known by the subject, on the one hand, and communicative reason, on the other." Again the "abyss" opens up in relation to Habermas's claim that modernity progresses by developing specialized systems of argumentation. Habermas holds discursive processes testing truths (about objects) and legitimacy (of social practices) to be entirely distinct. But the whole point of Latour's critique of "the modern constitution" is to show that "hybrids" breed in the space between a scientifically objectified nature and a political order instituted by free human subjects. It is to fill in that space that Latour proposes a legislative forum in which factual, scientific questions (e.g., "How will genetically modified strawberries behave when grown in open fields?") and questions about the extent of legitimate regulatory activity (e.g., "What is a reasonable level of precaution to impose on such experimentation?" or "Should companies be allowed to patent new genes?") are posed simultaneously. Latour envisions a deliberative process in which every constituent of a hybrid-forming network gets a chance to challenge others' views of the facts and to probe the appropriateness of their research methods, the reliability of their measuring instruments, and the moral acceptability of their political and economic connections.

For Duclos, it does not matter that Habermas locates reason in collective discursive practices rather than in the autonomously legislating subject. Either way, the diversity of our passions bars their rationalization.

Targeting Habermas. Duclos finds only illusion in the idea of a "super-discourse" above the passions, a discourse able to institute a stable order among them all. In practice, such a discourse would amount simply to the dominance of one passion over all the others—in particular, a passion for juridical normalization that can end up suppressing individuality (Duclos 1996: 183, 190). To avoid the obsessional application of any passion, Duclos says, we must learn to balance various forms of disengagement, not to overcome contradiction through reason. The task of a civil ecologism is to "re-establish an agreement around a 'nature' that is as little pre-defined as possible" (ibid.: 183).

Thus, there are features of French ecologism that cannot be reduced to the terms of discourse ethics. But this is not simply a philosophical stalemate. I want to suggest that these irreducible features could help strengthen a green discourse ethics at one of its weakest points. Just as French ecologists often presuppose (but do not articulate) the notion of progressively developing rational consensus, Habermas presupposes (but does not elaborate theoretically) a notion of divided natures.

Habermas makes a highly original contribution to critical theory—and that is where the problem begins. His whole pragmatic edifice depends on distinguishing two modes of human action, one grounded in work and the other in intersubjective communication (Whitebook 1996). The first is defined in relation to "external nature." Work is an expression of our technical interest in mastering the physical necessities of life through their prediction and control. This mode of action eventually gets formalized in the natural sciences. By objectifying the world, our species engages a "self-formative process" through which we learn to free ourselves from the blind forces of nature (Habermas 1971: 197).

In this conception of external nature alone, there is already a puzzling ambiguity. Building on the tradition of transcendental philosophy, Habermas contends that, through labor, our practical interest in controlling the conditions of our existence *constitutes* the only nature we can know: nature-for-us. This is what we experience as objective nature. Yet Habermas's roots in historical materialism lead him to regard the evolution of the human species as a "natural" process. Nature in this sense is something that neither human will nor social practice can alter. It is nature-in-itself. The puzzle created by this division is that it involves making a knowledge claim that Habermas's own theory rules out. If the only nature we can know is constituted by our pragmatic engagements with the world, then we have no grounds for saying that human interests arose in the "natural" history of the human species, for that nature is nature-in-itself (Vogel 1995: 29–30).

In addition to these two interpretations of external nature, Habermas's critical project demands a third. He refers to "internal nature"—that is, to the instinctual organization of the human psyche. He argues that we overcome the non-intentional, "natural" causality of unconscious motives and distorted perceptions by subjecting them to deliberative examination. If we are to escape illusion and neurosis, our sensations, needs, and feelings must be integrated into structures of linguistic interaction where they can be generalized and evaluated (McCarthy 1978: 315). We do not control inner nature by treating our psyches as technically manipulable objects. Dialogue in which people both define their separate identities and coordinate their intentions with others is the condition of a communicative rationality that masters natural compulsions.

In summary: Habermas invokes three "natures," each with different functions in his theory. Objective nature is the complement of our practical interest in freeing ourselves from the arbitrary forces of physical existence. Inner, subjective nature is the complement of our need for individuation and social integration. And nature-in-itself, as Joel Whitebook says (1996: 293), is "a theoretical posit which must be made to indicate the externality, contingency, and facticity of nature, which conspire to confound any arbitrary interpretations we seek to put on it."

Nonanthropocentrists will maintain that none of these natures is sufficient to allow Habermas to escape anthropocentrism. Nature-in-itself is unknowable and so cannot play a role in setting an ecological agenda. But then Habermas's view of nature-for-us makes it understandable only as an object of possible technical control in relation to human material needs (Vogel 1995: 298). And inner nature leads straight back to a human-oriented discourse ethics. If our inner natures make us seek to gratify our desires in mutually incompatible ways (for some, consuming the world; for others, simply admiring its beauty), then we can coordinate our actions only by seeking a rational consensus on generalizable forms of gratification. Only human interests count. Rational consensus provides no guarantees for the survival of the natures that are not readily interpretable as matters of human interests.

Some green theorists who are generally sympathetic to Habermas have taken these criticisms to heart. For them, three natures are not enough. Henning Ottman concludes that Habermas's lifeless, freely exploitable nature-for-us is really only nature-for-us-*moderns*—i.e., that it arose along with industrial-technical culture in the West. Today, however, increasingly frequent experiences of environmental disruption are overturning that nature. The new nature "reveals itself to be a *purpose-for-itself*, in the face of which the will-to-control has to impose limitations on itself" (Ottman 1982: 89). Ottman seeks a communicative ethics that finds a "compromise" between a nature that we control and one that transcends control.

John Dryzek argues that the idea of communicative rationality should be made to encompass "agency" in nature. True, nonhuman entities have neither language nor self-awareness. Still, they communicate in the sense of sending signals (e.g., the gestures of animals, chemical messages sent between plants, signs of distress in an ecosystem) to which we can pay attention and even accord respect. Whereas liberal democracy depends on aggregating human preferences, *deliberative* democracy defines itself in terms of undistorted communication. An ecologically amended Habermasian perspective allows us to imagine institutional designs that are especially adept at interpreting nature's signals accurately (Dryzek 1995: 24). Ecoanarchist ideas for communities scaled to the scope of ecological problems and proposals for invigorating civil society could inspire such designs.

Whether discourse ethics really can take on these amendments is questionable. Much depends on further elaboration of the ideas of agency and purpose that they invoke. Even if one grants the debatable assertion that the language of purposes is appropriate in describing the end-orientation of nonhuman things, surely the ethical implications of running up against those purposes cannot be same as encountering the purposes of human beings. In a communicative ethics, clashing human purposes get subjected to a deliberative process in which all parties involved seek to justify their action in universal, principled terms. Nonhuman purposes (e.g., homeostatic processes in an alpine ecosystem that maintain a constant pH in lake waters) are not set forth as truth claims or normative principles, so how can they become part of a process of discursive testing that might redeem or reject their "claims"?

Moreover, if signal-emitting entities cannot engage in the give and take of discursive testing, how are humans supposed to treat their "purposes"? Certainly we cannot grant them automatic deference. Even in the context of interhuman relations, we do not just defer to others' purposes. We bend our objectives to make them fit into a legitimate framework of action with others. But nature's "purposes" are inflexible, dogmatic, and uncompromising. They are what they are. Who can "negotiate" with such a partner? Anyway, automatic deference would utterly incapacitate human action, since nature's purposes are everywhere. If "compromise" between nature-in-itself and nature-for-us is advisable, it is nothing like compromise with another purposive being. Most likely it amounts to an anthropocentrically motivated adjustment in our relation to external nature.

Dryzek's "rescue" of communicative ethics from Habermas seems more promising. In a declaration that is unusual in the world of English-speaking ecologism, Dryzek (1995: 18) explicitly states his intention to "downplay 'centrism' of any kind." Yet there is reason to suspect that his rescue mission slips deeper into centered territory than he wishes. The idea of "respecting" signals emanating from the nonhuman world seems to suppose that we can distinguish real signals from the noise of the world. Before deciding whether Dryzek's position is tenable, we would need know how to differentiate between, say, an animal's cry of pain and light waves emitted when an atom's electrons move from a higher to a lower state of energy. Surely the latter is not a signal in an ethically relevant sense. If what makes a signal ethically relevant is the *life* of the signaler, we seem to be back on the path to biocentrism. Then again, Dryzek also speaks of signals from ecosystems. Ecosystems contain life, but it would be highly controversial to say that they have their *own* life. If excited atoms do not communicate with us in any ethically relevant sense, how can ecosystems be said to do so? One might try to work up criteria for identifying all morally considerable entities, as ecocentrists do. But the point was to downplay centrism.

There is no easy way out of these conundrums, and I do not mean to imply that French ecologists have found one. But I do think that their perspectives offer some decidedly noncentered strategies that could reinforce a green discourse ethic.

The study of French ecologism suggests that the very need to multiply natures is a clue that "nature" fills a number of incompatible roles in our understandings of our selves, our communities, and the inexhaustibly rich context in which both exist. Nature is not an agent in any ordinary sense. Nor can it be reduced to a collection of things, however complex their inter-relations. In its most general sense, nature is what a human society meets at the "limits of its agency and autonomy" (Hazelrigg 1995: 160). That definition suggests that, necessarily, nature both correlates to our practical activities and lies beyond them. Thinking of nature as something that tran-scends our control—as that which resists systematization, as whatever con-founds our expectations—gives discourse ethics a vitally needed addition: a skeptical dimension. It is in this dimension that French ecologists are at their strongest.

French Ecologism and the Tradition of Skeptical Humanism

With Montaigne, Pascal, Rousseau, and Voltaire in its philosophical pan-theon, France boasts a deeply rooted tradition of skeptical humanism. Although profoundly different in temperament, all these thinkers took full advantage of rhetorical resources that rationalists avoid. To challenge competing arguments, they traded on ambiguity, incompleteness, and per-ceptual shifting. Montaigne and Voltaire were profoundly distressed by political intolerance and fanaticism. Unlike liberal rationalists, however, they were not convinced that the progress of principles could vanquish those ills. Skeptics suspect that new ideas of individual rights or more compelling theories of justice do not necessarily instill in people a wari-ness of power. Often, indeed, it is *confidence* in the justice of their cause that blinds people to the suffering they inflict on others (Whiteside 1999: 501–508).

Thus, skeptics achieve ethical effects indirectly by questioning that con-fidence. They challenge epistemological monisms (e.g., rationalism or mate-rialism), insisting that reason is unable to establish uncontestable first principles. That is how Pascal found a place not only for science but also for instinct and religious faith in humanity's search for knowledge. Typically, skeptics also undermine such moral monisms as utilitarianism and Kantian ethics. They play on the irreducible diversity of values, and, like Rousseau, they decry the smugness with which elites assume that their values are the "natural" ones. Looking inward, skeptics cultivate an awareness of the complexities of the human heart. Like Montaigne, they see within themselves not so much a core or a universal nature as a multitude of contradictory dispositions that constitute their individuality.

French ecologists are skeptics in all these ways. They are epistemological pluralists who deny that scientific or economic experts enjoy special prerogatives in defining the understanding of nature that should guide political ecology. That was Serres's argument: that our understandings of the physical world have been shaped—and rightly so—by confrontations between scientists and the city in civil tribunals. A critique of economic expertise underlies Saint Marc's denunciation of the "ultraliberal" marketplace. Gorz and Charbonneau distrust accumulated power—including that of the liberal democratic state. As a skeptic, Duclos sees potential for cruelty and exclusion even in the supposedly most rational arguments for the social contract. When evaluating the rightness of social practices, skeptics insist on listening to the testimony of all who feel concerned or victimized—all the participants in Latour's parliament of things, for example—not just to accredited specialists and pre-legitimized representatives of the law. One answer to a rationalist's demand for an autonomous, rule-creating, logic-wielding center of evaluation is that what we are (as subjects), and what nature is in relation to us, vastly surpass the ken of any rational knower. In skeptical recognition of that fact, we learn the virtues of restraint, tolerance, and genuine reverence for life.

At an institutional level, skeptics are too cautious to offer social blueprints. But they are not completely at a loss for ideas. Morin opines that an ecological society would require an "acentric organization" whose virtue is to leave room for uncertainty and disorder. Some disorder allows a society to be receptive to sources of creativity (Morin 1980: 326–328). This is an important clue to the skeptic's social strategy. Skeptics seek to *reposition* us so that we see the humanity-nature nexus differently and have new opportunities to act on our perceptions. Deléage's hopes for organizations that develop environmental "counter-expertise" clearly fall under this description.

So too does Latour's parliament of things. This parliament is best understood, I think, not as a particular legislative chamber but as a multiplication of sites throughout a society where citizens can challenge every stage in the process by which "hybrids" are created and disseminated (Latour 1999b: 222). For Latour (ibid.: 99), disagreements among scientists over the reality of global warming and other ecological problems motivate the first skeptical lesson for participants in the parliament: "to know how to doubt" those who speak on behalf of things. From the personalists' regionalism to the ecosocialists' advocacy of new social movements, the French ecologists' proposals lean toward decentralized political structures and vigorously contestative nongovernmental bodies. But the purpose of decentralization as envisaged by French ecologists is not to pick up on "feedback signals emanating from natural systems" (Dryzek 1995: 24). French proposals aim to keep in the foreground the contestable social and knowledge-forming processes through which communities develop their conceptions of nature.

Such institutional ideas, one might say, are designed to be particularly sensitive to expressions of *doubt*: doubt about risk assessments, about expert reassurances that pollution levels are safe, and about claims that goods always better serve society when they are exchangeable in a regime of property rights. Duclos showed why turning our ear to those human expressions counts as political ecology. Behind people's doubts are divided natures—the scientists' objectified nature, the pre-political nature from which we escape through the social contract, the inner nature of instinctual pleasures. These natures assert themselves against contemporary trends that tend to homogenize values and behavior. That is why seemingly strictly "human" issues (e.g., rampant commercialism and workplace hierarchies) are integral to green practice. Our approaches to those "social" problems turn on conceptions of "nature" that are inseparable from the conceptions of nature through which we frame more obviously environmental issues. Institutions that give these doubts a hearing not only counteract what Habermas calls the "one-sided" development of modernity; they also protect the divisions that keep our multiple "natures" alive.

These ideas, if kept within measure, do not contradict a discourse ethic. They supply it with an essential complement. Communicative ethics needs a skeptical dimension because no rationality can bring order to all our moral conflicts and truth to all our understandings of nature. What happens when a claim about legitimacy spills out of the moral-legal realm and runs into the domain of scientific inquiry? That is just what happens when greens charge that our understanding of the potential dangers of genetically modified crops is skewed by the influence of agribusiness money on programs of university research. What happens when an aesthetic claim collides with a moral one? That is just what happens when a personalist ecologist asserts that standing awestruck before a natural scene develops in the individual a type of "wild" freedom that a just society has no right to repress. Though each of modernity's cultural spheres can be rationalized on its own terms, there is no higher logic that settles conflicts between validity claims that cross the spheres. In the final analysis, according to White (1988: 135), Habermas admits that differently institutionalized forms of argument can only be "balanced." However, this implies that—in even the most rational society—actors must be able to move between discourses without being able to explain fully why they do so.

That is why Duclos (1993c) calls for "a new definition of nature . . . that can ground 'non-action,' . . . restraint, respect for discontinuities, free spaces, differences." In effect, skepticism offers green theorists a way to rethink the very notion of "ecological limits." Centered thinkers conceive of limits as scientifically verified thresholds of ecosystemic health or as respectable identities of nonhuman things. Skeptics, in contrast, suggest that avoidance of environmental destruction depends not on observing precise boundaries but on practicing considered restraint.

The skeptical orientation of noncentered ecologism demands that a theory of universal pragmatics face up to its own peculiar danger: the danger of pushing for action beyond the limits of its own justification. Communicative ethics is action oriented to the core. The very purpose of discursive testing is to remove distortions in communication that impede our ability to coordinate our actions with those of others. But if we must "balance" various spheres of communicative action, there must be instances when it would be wrong to move ahead simply because action had been fully justified in one sphere. And that is Duclos's argument: There must be instances of "non-action"—spaces where we remain free from the demands of social coordination. Skeptics urge us to see that there are times when we must "let beings be" out of recognition that our cognitive and ethical schemata, while enabling our action, incite us unwittingly to destroy what they cannot encompass.

Yes, skepticism can be pushed too far. In this, I think, Habermas is right. Skepticism can become a refusal to argue. Such a refusal, Habermas contends (1990: 86), can only be an "empty gesture," because the skeptic "cannot drop out of the communicative practice of everyday life" and its "presuppositions of argumentation." French ecologists, by and large, do not extend skepticism to the point of undermining all knowledge. Not one of the theorists discussed in this book disregards ecological science. None proposes simply to abolish markets and their attendant logic of economic efficiency, and each assumes that democracy legitimizes community activity better than any alternative. Moreover, each of these theorists, in taking a noncentered perspective, actually hopes that his own writings can make us conscious of ecologically beneficial opportunities that, without his theory, we would miss. Noncentered ecologists, in other words, do aim to alter our action orientations through argument—even if the alteration is sometimes to stop us in our tracks.

Conclusion

Although skepticism cannot replace the other rationalizing modes of argument, it has a special function that justifies keeping it among the rhetorical resources of a discursive ethics. To demonstrate this, I call on noncentered ecologism and Habermasian theory to elucidate each other one last time. Noncentered ecologists evoke an ideal of multicompetent individuality that is embedded in Habermasian social theory.

Discourse ethics rests on a conception of communicative competence. For Habermas, what makes a person communicatively competent is an ability to engage in discursive practices that create a shared world. Communicatively competent individuals use speech acts to reach understanding with others. But the theory of how modernity develops specialized spheres of argumentation seems to imply that, in a society with a complex division of labor, individuals will probably become especially competent in only one sphere. Natural scientists learn highly developed protocols for evaluating truth claims in their field of specialization. Other citizens become especially skilled at subjecting proposed regulative norms to tests of fairness. Meanwhile, poets and artists may understand better than most the array of aesthetic modes through which subjectivity can gain authentic expression. Habermas postulates that evolution toward a rational society depends on the distribution of these competences in the social system as a whole. He does not insist, however, on developing all the communicative competences in any particular individual. Yet that is exactly what his notion of "balancing" validity claims requires. Institutional balancing is not enough, because institutions develop internal cultures that reward exclusive conceptions of validity. One hears the consequences of such exclusive acculturation in everyday environmental debates. There is testimony from natural scientists who deplore the public's lack of sophistication in evaluating the risks of new technologies—as if only truth claims were at issue. There are economists who believe that all values are monetizable, just as there are jurists who declare the state's pre-eminence in every potential dispute. Poets, too, have been known to weigh in with far-reaching ideas of their own. Gary Snyder (1974: 23) stings us with the sense of loss we will have once industrial civilization has left "no place / A Coyote could hide." His poetic imagery evokes a preservationist claim, however indeterminate.

When these logics collide in public debate, what can "balancing" them mean? Who has the standing to declare that an outcome is balanced? Balance must signify more than whatever happens to emerge from the interaction of forces in the deliberative arena. It implies that there are people who are able to perceive disproportion in the way that various rationalities get applied in society as a whole. If imbalance is to be arrested, there must be some actors who are aware that claims emerging from one sphere or another are overreaching. A decision to counter such overreaching cannot result from the argumentative force of a superior validity claim. Whether any claim is superior is precisely what is at issue. The decision depends on *someone's* sense that, at times, the wisest course is to draw back from any final, action-enabling conclusion.

Provoking that wisdom is the skeptic's special communicative competence. It would be a great error to believe that the skeptic's doubts signal a deficiency of knowledge. Quite the contrary. To carry conviction, the skeptic must know enough about argument in each sphere to be able to detect reason overextending itself. The skeptic's skill involves calling attention to what is unknown, paradoxical, repressed, contradictory, unintentional, ineffable. Skeptics perceive opportunities to set one logic against another precisely because they understand each logic well enough to see important aspects of reality that escape its grasp. Taking advantage of such opportunities is a matter of having an exceptional facility at deploying knowledge across the spheres. Skeptics are intrepid travelers who dare to move among worlds as different as those of the risk-assessment specialist, the poet, and the molecular biologist. The traveling does not make them doubt all communication. Like seasoned translators, they believe that much meaning can pass between differing communities of discourse. But their experience moving between languages tells them also that some nuances of meaning inevitably escape translation. Skeptics try to find the language to convey to each community why the other expresses some—*only* some—of the truth. The only way to respect that fact is to cultivate an appreciation of the limits of our social logics. This appreciation is not the negation of rationality but rather an essential condition of it. Noncentered ecologism challenges us to develop a paradoxical ideal: an ideal of skeptical expertise.

That ideal helps explain, I believe, one of the most beguiling but also one of the most mysterious ideas encountered in French ecologism. Michel Serres's work on the natural contract (1991) culminates in praise of a figure whom he calls the *tiers-instruit*—the "educated third." Serres (1990: 147–148) envisions a person who is "expert of cognizance, formal or experimental, versed in the natural sciences of the living and nonliving . . . , traveler of nature and society, lover of rivers, sand, winds, seas and mountains . . . , solitary sailor across the North-West Strait, latitudes where positive mixed knowledge communicates in a delicate and rare manner with humanities . . . , more distant from power than every possible legislator . . . , capable of recognizing and understanding failures . . . , and finally, and above all, burning with love for the earth and humanity." The "third's" distance from power suggests a capacity to make judgments that do not fall wholly under the logic of legitimization. At the same time, her understanding of "positive mixed knowledges" and the humanities gives her perspective on every point where "knowledge communicates"—and where communication fails. She is more familiar with the natural sciences than most humanists, more deliberative and more cognizant of the cultural inflections of nature than most scientists, and (thanks to the human sciences) more skeptical about social dogma than political activists have been in the past (Serres 1992a: 257–265).

As our technological power increases, everything—landscapes, oceans, species, the temperature of the earth, our own genetic identities—becomes subject to our decisions. If such power is not to end in wanton destruction, "we are going to need," says Serres (1992b), "an immense wisdom, a wisdom proportionate to this prodigious knowledge." That wisdom demands far more than the mastery of any specialized discipline, and it cannot be obtained through contemplation alone. It demands a discipline of transdisciplinarity. We need natural scientists who also really understand the claims of distributive justice. We need economists who also respect notions of incommensurability and beauty beyond price. We need jurists and political activists who also keep sufficiently abreast of biology to follow the basic dynamics of species interactions in ecosystems. And we need connoisseurs of literature and philosophy whose convictions about the meaning of life find expression in a language that touches the souls of technicians and policy analysts.

French ecologism teaches that green politics demands nothing less than a new humanism. Such an ethic announces neither human superiority over nature nor humanity's subordination to nature. A new ecological humanism provokes deliberation about all the natures that divide us. In that project, immense wisdom dons the guise of the skeptic's modesty.

Notes

Introduction

1. Impressively comprehensive English-language surveys of green political now abound. See Dobson 1995; Eckersley 1992; Atkinson 1991; Davis 1989; Baxter 1999; Hayward 1994. All of them omit French ecologists, with the exception of André Gorz.

2. See Lecourt 1997: 23.

3. See also Pepper 1984: 52; Taylor 1986: 143; Mathews 1991: 32.

4. French popular opinion has become more favorable to green perspectives since the 1970s. See Kalaora 1998: 163–189.

5. Some writers (e.g. Dobson 1995: 1–8; Faucher 1999: 37) distinguish environmentalism (which usually implies resource management and conservation) from ecologism and green thought (which demand farther-reaching changes in production and consumption to create a sustainable relationship between humanity and the nonhuman world). Other writers handle that controversial distinction quite differently. I use 'environmental,' 'ecological', and 'green' interchangeably.

6. Ferry analyzes only two French theorists of ecologism: Michel Serres and Félix Guattari. He seriously misinterprets Serres (see chapter 4), and he fails to implicate Guattari in the spread of nonanthropocentrism that supposedly is the dangerous perspective behind the "new ecological order."

7. Quoted in Dallmayr 1987.

8. John Barry (1999a: 99–104; 1999b: 126) furnishes material for both potential objections. His own project of "rethinking green theory," which relies heavily on British sources, provides grounds for saying that British ecologism, like French ecologism, is less preoccupied with wilderness preservation and more constantly concerned with social justice than some New World species of English-speaking ecologism. But the very need he feels to criticize the nonanthropocentric arguments of various English-speaking theorists, and his lack of attention to any French thinker, show that the rhetorical field of English-speaking ecologism is still very much in evidence. (It is in the present book, too!) France has an especially strong reputation as "a country that magnifies artifice" (Jacob 1999: 310), whereas Germany (Cans

1997: 210) and the Netherlands (Keulartz 1999: 87–91), whose territories have been humanized at least as much as France's, nonetheless have spawned ecological thinkers who are determined proponents of "wilderness." The virtual absence of such thinking among prominent French ecologists alerts us to something distinctive about the French context.

9. For historically detailed accounts of the French ecology movement and the swirl of ideas surrounding it, see Sainteny 2000; Jacob 1999; Shull 1999; Prendiville 1993.

Chapter 1

1. Of course, antecedents of what might today be called "environmental protection" initiatives occurred long before the nineteenth century. Henry IV ordered the construction of sewers in Paris in the sixteenth century. Under Louis XIV, Colbert issued decrees to protect royal forests as sources of timber for the navy (Stevens 1991: 25; Duclos and Smadja 1985: 135–136; Larrère 1997: 90–91). In France, as elsewhere, ecological thinking emerged in the middle of the nineteenth century out of a confluence of disciplines such as geography, botany, and "natural history." On the history of ecological science in France, see Drouin 1991. Although Saint-Hilaire is frequently discussed in this context, another French contributor to the discipline also deserves mention: the anarchist and geographer Élisée Reclus, author of *L'Homme et la terre* (1866). See Chesneaux 1993.

2. All the quotations in this paragraph are from Jacob 1999 (part I, chapter 2). According to Jacob, Hainard's "conservative naturalism" is comparable to Moscovici's "subversive naturalism" in founding a major strain of French ecologism. See Jacob 1999: 8, 115–160, 311. My own research does not confirm this view. Of all the figures reviewed in this work, only Waechter and Lebreton claim some inspiration from Hainard.

3. Quotations from Jouvenel's writings of the 1950s can be found in Launay 1997.

4. Later, Jouvenel would be a member of the Club of Rome, whose report on *The Limits of Growth* did so much to stimulate environmental concern in the 1970s. His works *Arcadie* (1970) and *La civilisation de la puissance* [*The Civilization of Power*] (1976) extend his critique of productivism and materialism (Jacob 1999: 212–216).

5. Today as in 1978, Lebreton is a professor of biology and the environment at Scientific and Medical University of Lyon. His stature as a scientific ecologist and his credentials as an environmental activist (for example, in opposing development of the Vanoise national park), as well as the influence of ideas developed in his book *L'Ex-croissance*—a pun on "croissance" (growth) and "excresence"—made him a rising star in the ecology movement in the 1970s. He has been less prominent since 1980, when, in a close primary, Brice Lalonde beat him out to represent ecologists in the 1981 presidential election (Pronier and Le Seigneur, 1992: 61; Jacob 1999: 140, 146).

6. See Boy 1991, p. 38.

7. For a more detailed discussion of Dumont's role, see Whiteside 1997.

8. On how Les Verts used Waechter's themes to distinguish themselves from their closest ideological rival, the French Socialist Party, see Shull 1999.

9. See also Sainteny 1991: 53–55; Whiteside 1994: 340–344.

10. Compare Dumont 1988: 56–74.

11. For a sociological analysis of some of the these experiments, see Léger and Hervieu 1979: 37–68.

12. Occasional exceptions exist. John Barry's *Environment and Social Theory* begins by pointing out ambiguities in the concept of nature. We do better, he says, to remain attentive to the "social and cultural meanings attached to the environment" (1999: 31). *Contested Natures* (1998) by Macnaghten and Urry, is even closer in spirit to noncentered ecologism.

Chapter 2

1. See Nash 1982; Taylor 1992: 1–15.

2. In Australia, English-speaking European settlers perceived their environment in much the same terms as North Americans have theirs: at times as "gloomy" woods unbroken by civilization, as "wastelands" suitable for settlement, or in pastoral-Romantic terms as exotic landscape. Although aboriginal peoples, over thousands of years, deliberately modified the landscape by the use of fire, "the landscape [in the eyes of the original settlers] bore no direct evidence that it was used, made productive or converted from its primeval state." See Frawley 1994.

3. For an engaging discussion of these transformations that is attentive to the influence of soil quality and microclimates on human settlements throughout France, see Braudel 1989.

4. In the 1970s Moscovici joined Friends of the Earth, then under the leadership of Brice Lalonde (Jacob 1995: 126).

5. Following Bernard Williams, I use the term "analytical tradition" broadly. It refers to a style of philosophizing that, while closely identified with early twentieth-century linguistic philosophy (Bertrand Russell, G. E. Moore), extends to contemporary ethicists such as John Rawls and R. M. Hare (Williams 1980).

6. Luc Ferry (1992: 54–55) argues that Kant's sensitivity to purposiveness in animals forbids instrumentalizing them *gratuitously*. But that concession is so weak that it barely responds to those who object to instrumentalizing nature. Ferry's "respect" for animals does not forbid our eating them or turning them into automobile upholstery. And since he sees purposiveness as a quality of an intending consciousness, his sympathy for animals says nothing about the moral status of species or ecological systems.

7. See also Taylor 1992: 117.

8. See also Taylor 1986: 135; Eckersley 1992: 2; Rolston 1988: 339; Baxter 1999: 57.

9. Not all French ecologists agree. Henri Chevallier (1982: 129) associates Rousseau with Jacobinism, not with the sorts of decentralization preferred in the ecology movement.

Chapter 3

1. In more recent years, Passet has been a professor at the Université de Paris I and the director of the Centre économie-espace-environnement.

2. Passet's writings in favor of a "multidimensional economy" (1989: 4–5) indicate that his position has not changed.

3. The volumes are (I) *La Nature de la Nature* (1977), (II) *La Vie de la Vie* (1980), (III) *La Connaissance de la Connaissance* (1986), and (IV) *Les Idées: Leur habitat, leur vie, leur moeurs, leur organisation* (1991).

4. "We must understand," Morin (1990: 257) writes, "that what I call method can be considered as a meta-method in relation to the scientific method. It does not cancel out scientific methods; on the contrary, it admits and recognizes them. But it questions, criticizes, checks and at times goes beyond scientific methods by its will to reflexivity."

5. See, e.g., Devall and Sessions 1985: 96–98.

6. Myron Kofman (1996: vii, 2) views Morin's work as "a revival of Montaigne's scepticism" and "the voice, if not the prophet, of a new humanism."

Chapter 4

1. See also Serres 2000: 12, 16–17.

2. For a useful overview in English of Serres's works up until 1980, see the introduction to Harari and Bell 1982. See also Mortley 1991.

3. Stengers is a periodic contributor to French ecologism. Latour (1999b: 321) sees affinities between his ideas and those in her *Cosmopolitiques—Tome 1, La guerre des sciences* (1996). See also Conley 1997: 68–75.

4. For more on Serres's unusual interpretation of cybernetics, see Conley 1997: 60–67.

5. See also Serres 2000: 7.

6. Larrère (1993: 45) says that Serres's theory would further empower not citizens, but scientists, since scientists are the ones who have to interpret the very "links and interactions" that make up nature's part of the contract. Not so. Serres (1990: 109–110) maintains that, since their activities affect the well-being of all, scientists and technicians should participate in, and be subject to, a renegotiated social contract.

7. For the connection between Serres and Latour, see Wesling 1997: 198.

8. For a critical review of Latour's early work, see Brown 1991.

9. Schaffer (1991: 182) accuses Latour of the heresy of hylozoism: "an attribution of purpose, will and life to inanimate matter, and of human interests to the non-human."

10. Works of English-speaking ecologists showing the marks of poststructuralism include Atkinson 1991; White 1998; Hayden 1997; Bennett and Chaloupka 1993; and, most notable, Conley 1997. Conley offers one of the rare discussions in English of French ecologists Michel Serres, Luc Ferry, and Félix Guattari. As such, her work forms a valuable complement to the present study. However, I disagree with her claim that "the driving force of poststructural thought is indissolubly linked to ecology" (Conley 1997: 7)—if that is meant to describe the significance of postmodern ecologism in France. Some of the thinkers she examines (e.g., Derrida, Lyotard) have written only minimally on ecology. Paul Virilio consistently uses the word "ecology" (e.g. 1978), but his ruminations on the effects of information and transportation technologies on human consciousness barely engage with concerns over "nature."

Chapter 5

1. Quoted on p. 133 of Prendiville 1993. Original source: *Le Personnalisme* (1903).

2. For overviews of pre-World War II personalism, see Loubet del Bayle 1987 and Rauch 1972.

3. Charbonneau's words in this paragraph appeared originally in "Le sentiment de la nature, force révolutionnaire," *Journal Intérieur des groupes personnalistes du Sud-Ouest* (June 1937), and are quoted in Roy 1997: 43–44.

4. Ellul's critique of technological imperialism in *La technique ou l'enjeu du siècle* (1954) undoubtedly feeds into French ecologism. See Guillebaud 1992 and Jacob 1994: 100–102. But Ellul rarely applied his critique explicitly to environmental issues. On Ellul as a "precursor" of political ecology in France, see Troude-Chastenet 1998.

5. The term "Gascon personalism" comes from Gascony, an old province of France, centered around Bordeaux. On Charbonneau, Ellul, and Gascony, see Roy 1997: 35–49.

6. French ecological activists, including Antoine Waechter, regularly mention Rougement in their list of formative intellectual influences. See Jacob 1994: 150, 155, and Journès 1979: 248.

7. The environmental concern of Rougement's wife, Nanik, was apparently decisive in causing this turn in his thought (Saint-Ouen 1995: 32).

8. See also Berg and Tukel 1980.

9. See also Saint-Ouen 1995: 30–31.

10. The charter appears in Vadrot 1978: 75–80. Vadrot (ibid.: 177) characterizes this charter as a bland, pleasing-to-everyone program that was mainly an attempt by centrists to set themselves up as interlocutors with the new Ministry of the Environment.

11. See also Charbonneau 1980: 99. Saint Marc (1994: 372–373) proudly defends his endeavors.

12. Saint Marc quotes the Bible regularly and devoutly invokes "history, which reaches its completion in God" (1978: 200; cf. 1971: 98).

13. In the words of the French theologians Hélène and Jean Bastaire (1996: 69): "Safeguarding creation begins with an internal decentering that . . . leads each person to stop seeing reality from his own point of view in order to see it from God's."

14. Source: interview by K. Whiteside, March 3, 1999.

15. Even when Lacan says that the "law" of language superimposes "the kingdom of culture on that of nature," "nature" cannot mean what it does for Duclos, something "surpassing the determinations of human language." Compare Lacan 1977: 66 and Duclos 1993a: 67–68. For Lacan, "reality" is "what resists symbolization absolutely." Resistance comes not from a physical world, but from the "dead letters" of signification that are sedimented in language (Lacan 1988: 66).

Chapter 6

1. Ecofeminism, explains Robyn Eckersley, "exposes and celebrates what has traditionally been regarded as other—both woman and nonhuman nature—in the context of a far-reaching critique of hierarchical dualism and 'masculine culture'" (1992: 64). Although feminist sympathies clearly show through in the work of many French ecologists, the absence of a strong current of ecofeminist theory in France is somewhat puzzling. None of the standard surveys of ecological thought in France (mentioned in chapter 1) treats ecofeminism as a significant element in green ideology. Luc Ferry at least mentions what makes this omission so odd. A French woman, Françoise d'Eaubonne, originated the term "ecofeminism" (1974). In attacking ecofeminism, however, Ferry immediately moves on to its English-speaking proponents such as Karen Warren and Carolyn Merchant (1992: 220–236). Conley is to be recommended for highlighting ecological themes in the works of Hélène Cixous and Luce Irigaray (1997: 123–140). Even Conley grants, however, that their work is so lacking in a practical dimension that it risks being "consigned either to history or to a lower rung on the ladder of ecofeminism" (1997: 140).

2. For further discussion of the idea of the body-subject, see Whiteside 1988: 22, 50.

3. Duclos's (1991b: 15–16) research on environmental and safety issues at the factory level suggests that workers are hardly immune to tendencies to defend their jobs by minimizing workplace dangers.

4. Source: interview with Jean-Paul Deléage, April 22, 1991.

5. Here and in chapter 8, I draw on some of my earlier writings on Lipietz. See Whiteside 1996.

6. Frederick Taylor's *Principles of Scientific Management* (1911) spelled out methods to increase the productivity of laborers, including separating those who design

production processes from those who execute them, implementing time-motion studies of workers, simplifying and standardizing production routines.

7. The following analysis of the crisis of Fordism summarizes pp. 17–58 of Lipietz 1989.

8. This is not simply American-style pluralism, whose observers too often play down the importance of money, education, and personal contacts in groups' access to power. The "interests" that Lipietz most wants to see organized are those that such "pluralism" most often disfavors: the poor, the socially excluded, the environmentally endangered.

Chapter 7

1. Translating "liberal" and related ideological terms is a task fraught with potential confusion. Ferry and others whom English-speakers would call "liberals" (i.e., theorists of human rights and limited, representative government) often avoid talk of "liberalism," preferring "democracy" instead. "Libéralisme"—in the eyes of French critics, at least—is often equated with advocacy of free markets. Meanwhile, French *libertaires* (libertarians) usually see themselves in the anarchist tradition. They are critics of pervasive market relations, not (like American libertarians) their most ardent advocates.

2. See also Grosser 1991 and Allègre 1990: 377.

3. Ophuls tries to mitigate this conclusion in his revised edition by calling for a transformation in individual values to favor steady-state principles.

4. See Alphandery et al.: 1991: 198; Theys 1993: 63; Duclos 1996: 4; Oudin 1996: 209.

5. Bourg's survey of ecologism does have the merit of discussing certain forms of "democratic ecology" (represented by André Gorz, Ivan Illich, and Philippe van Parijs) that Ferry ignores.

6. Thomas (1992: 165–173) asserts a filiation running from Barrès's *Les Déracinés* [*The Uprooted*] ("starting in 1897, the bible of young people torn away from the country") to Pétain (who declared "The earth does not lie") to the ideas of a couple of contemporary greens who clumsily defended regionalism. For a more sober evaluation of such a lineage, see Alphandéry et al. 1991a: 243–249.

7. Tazieff is a well-known vulcanologist and former French Secretary of State for the Prevention of Major Risks (Cans 1991). A critic of those who predict ecological catastrophes (Tazieff 1992), he has also done much to publicize environmental problems in France.

8. There is nothing uniquely ecosocialist about Lipietz's arguments in this regard. For the same ideas cast in liberal terms, see Wells and Lynch 2000: 106 ff.

9. For a time, Allègre was also Minister of Education in Lionel Jospin's socialist-green coalition government.

10. Oudin (1996: 24–29) doubts that chlorofluorocarbons are causing the "ozone hole." Lenoir (1992) minimizes the likelihood that human activities are provoking global warming.

11. The liberal Olivier Postel-Vinay (1998) confirms Duclos's suspicion. Postel-Vinay recommends the principle of precaution because it demands that "scientific experts" and "economic experts" meet for a "healthy discussion" in which "common sense" can prevail over "fundamentalisms" (i.e., either green or libertarian views).

12. Ferry's (1992: 216–217) criticism of Guattari's notion of a democratic right to "dissensus" foreshadows such a charge but does not apply it widely.

Chapter 8

1. In fact, in his 1999 book *Qu'est-ce que l'écologie politique?* Lipietz suggests for the first time that Habermasian discursive ethics offers one of the two "respectable" paths that lead to political ecology. The other, he argues (p. 33), "is rooted in the great religions, embodied in our day by Emmanuel Lévinas or Hans Jonas." Lipietz does not choose between these two paths.

Bibliography

Abélès, Marc, ed. 1993. *Le Défi écologiste*. L'Harmattan.

Acot, Pascal. 1988. *Histoire de l'écologie*. Presses Universitaires de France.

Alexander, Don. 1990. Bioregionalism: Science or sensibility? *Environmental Ethics* 12: 161–173.

Allan Michaud, Dominique. 1989. *L'Avenir de la société alternative: Les idées 1968–1990*. L'Harmattan.

Allègre, Claude. 1990. *Économiser la planète*. Fayard.

Alphandéry, Pierre, Pierre Bitoun, and Yves Dupont. 1991. *L'équivoque écologique*. La Découverte.

Anderson, Brian. 1997. *Raymond Aron: The Recovery of the Political*. Rowman and Littlefield.

Anger, Didier. 1981. La formation d'une nouvelle sensibilité. *Que Faire Aujourd'hui?* 16: 35–36.

Ariane, Chemin, and Fabre Clarisse. 1997. Dominique Voynet, ministre autrement. *Le Monde*, December 15.

Atkinson, Adrian. 1991. *Principles of Political Ecology*. Bellhaven.

Atlan, Henri. 1972. *L'Organisation biologique et la théorie de l'information*. Hermann.

Barry, John. 1999a. *Environment and Social Theory*. Routledge.

Barry, John. 1999b. *Rethinking Green Politics*. Sage.

Bastaire, Hélène, and Jean Bastaire. 1996. *Le salut de la création: Essai d'écologie chrétienne*. Desclée de Brouwer.

Baudrillard, Jean. 1994. *The Illusion of the End*. Stanford University Press.

Baxter, Brian. 1999. *Ecologism: An Introduction*. Georgetown University Press.

Bennahmias, Jean-Luc, and Agnès Roche. 1992. *Des Verts de toutes les couleurs: Histoire et sociologie du mouvement écolo*. Albin Michel.

Bennett, Jane, and William Chaloupka, eds. 1993. *In the Nature of Things: Language, Politics and the Environment*. University of Minnesota Press.

Benton, Ted. 1993. *Natural Relations: Ecology, Animals and Social Justice*. Verso.

Benton, Ted. 1995. A green and pleasant land. *New Times*, January 21.

Berg, Peter. 1991. What is bioregionalism? *Trumpeter* 8: 6–8.

Berg, Peter, and George Tukel. 1980. *Renewable Energy and Bioregions: A New Context for Public Policy*. Planet Drum.

Besset, Jean-Paul. 1992. *René Dumont: Une vie saisie par l'écologie*. Stock.

Bianchi, Françoise. 1990. Lecture hologrammatique de l'oeuvre d'Edgar Morin. In *Arguments pour une méthode (Autour d'Edgar Morin)*, ed. D. Bougnoux et al. Seuil.

Bogue, Ronald. 1989. *Deleuze and Guattari*. Routledge.

Bookchin, Murray. 1982. *The Ecology of Freedom: The Emergence and Dissolution of Hierarchy*. Cheshire.

Bookchin, Murray. 1989. *Remaking Society*. South End.

Bookchin, Murray. 1990. *The Philosophy of Social Ecology: Essays on Dialectical Naturalism*, second edition (Black Rose, 1995).

Bourg, Dominique. 1992. Droits de l'homme et l'écologie. *Esprit*, October: 80–94.

Bourg, Dominique, ed. 1993. *Les Sentiments de la nature*. La Découverte.

Bourg, Dominique. 1994–95. Sciences, nature et modernité. *Écologie Politique* 11–12, winter: 113–136.

Bourg, Dominique. 1996a. *L'Homme artifice: Le sens de la technique*. Gallimard.

Bourg, Dominique. 1996b. *Les scénarios de l'écologie*. Hachette.

Bourg, Dominique. 1998. *Planète sous contrôle*. Textuel.

Bowring, Finn. 1995. André Gorz: Ecology, system and lifeworld. *Capitalism, Nature, Socialism* 6: 65–84.

Boy, Daniel. 1990. Comment devient-on un parti? *Politis* 9, first trimester: 15–17.

Boy, Daniel. 1991. Le vote écologiste: évolutions et structures. Cahiers du CEVIPOF no. 6.

Boy, Daniel, Vincent Jacques le Seigneur, and Agnès Roche. 1995. *L'Écologie au Pouvoir*. Presses de la Fondation Nationale des Sciences Politiques.

Boyer, Robert. 1990. *The Regulation School: A Critical Introduction*. Columbia University Press.

Bramoullé, Gérard. 1993a. Malthusiana: Critique du programme économique des Verts. *Krisis* 15, September: 30–36.

Bramoullé, Gérard, and Alain Lipietz. 1993b. Face à face: Libéralisme ou écologisme. *Krisis* 15, September: 37–50.

Braudel, Fernand. 1986. *The Identity of France*. HarperCollins.

Brodhag, Christian. 1990. *Objectif Terre: Les Verts, de l'écologie à la politique*. Félin.

Brodhag, Christian. 1994. *Les Quatre vérités de la planète: Pour une autre civilisation*. Félin.

Brown, James Robert. 1991. Latour's prosaic science. *Canadian Journal of Philosophy* 21: 245–261.

Brulle, Robert J. 2000. *Agency, Democracy, and Nature: The U.S. Environmental Movement from a Critical Theory Perspective*. MIT Press.

Buchmann, Andrée. 1990. Deux Verts en politique. *Politix* 9: 7–9.

Callicott, J. Baird. 1989. *In Defense of the Land Ethic*. State University of New York Press.

Cans, Roger. 1991. *La Passion de la Terre*. FIRST.

Cans, Roger. 1992. *Tous Verts! La surenchère écologique*. Calmann-Lévy.

Cans, Roger. 1997. Les trois soeurs de l'écologie. In *Environnement: Représentations et concepts de la nature*, ed. J.-M. Besse and I. Roussel. L'Harmattan.

Capra, Fritjof. 1982. *The Turning Point: Science, Society and the Rising Culture*. Simon & Schuster.

Carlier, Jean. 1994–95. Naissance et avatars de l'écologie politique. *Aménagement et Nature* 116, winter.

Chalanset, Alice. 1997. *Les sources de l'écologie*. Pleins Feux.

Champetier, Charles. 1993. Le vieil ordre moderne: À propos du 'Nouvel Ordre Écologique' de Luc Ferry. *Krisis* 15, September: 72–95.

Charbonneau, Bernard. 1951. *L'État* (Économica, 1987).

Charbonneau, Bernard. 1973. *Le Système et le chaos: Critique du développement exponentielle*. Anthropos.

Charbonneau, Bernard. 1980. *Le feu vert: Autocritique du mouvement écologique*. Karthala.

Charbonneau, Bernard. 1991a. *Nuit et jour: Science et culture*. Économica.

Charbonneau, Bernard. 1991b. *Sauver nos régions: écologie et sociétés locales*. Sang de la terre.

Charbonneau, Bernard. 1992. La révolution impossible et nécessaire. *Combat Nature*, November: 38–40.

Chesnaux, Jean. 1993. Du sentiment de la nature dans la société moderne: Elisée Reclus. Présentation par Jean Chesnaux. *Écologie Politique* 5, winter: 154–173.

Chevallier, Henri. 1982. *Éléments pour une écologie politique*. Ende Doman.

Conley, Verena Andermatt. 1997. *Ecopolitics: The Environment in Poststructuralist Theory*. Routledge.

Crahay, Anne. 1988. *Michel Serres, La mutation du cogito: Genèse du transcendental objectif*. Éditions Universitaires.

Dallmayr, Fred R. 1987. *Critical Encounters between Philosophy and Politics*. University of Notre Dame Press.

Daly, H. 1992. *Steady-State Economics*, second edition. Earthscan.

Davis, Donald Edward. 1989. *Ecophilosophy: A Field Guide to the Literature*. Miles.

Davis, Mary D. 1988. *The Ecologist's Guide to France.* Green Print.

d'Eaubonne, Françoise. 1974. Le féminisme ou la mort. Pierre Horay. ["The time for ecofeminism," in *Ecology*, ed. C. Merchant and R. Gottlieb. Humanities, 1994.]

Debeir, Jean-Claude; Jean-Paul Deléage, and Daniel Hémery. 1986. *Les Servitudes de la puissance: Une histoire de l'énergie.* Flammarion. [*In the Servitude of Power: Energy and Civilization through the Ages.* Zed, 1991.]

Deléage, Jean-Paul. 1992a. Écologie: Les nouvelles exigences théoriques. *Revue de l'Écologie Politique* 1: 1–12.

Deléage, Jean-Paul. 1992b. *Histoire de l'écologie: Une science de l'homme et de la nature.* La Découverte.

Deléage, Jean-Paul. 1993a. L'écologie, humanisme de notre temps. *Revue de l'Écologie Politique* 5: 1–14.

Deléage, Jean-Paul. 1993b. Au-delà de l'échec: L'avenir de l'écologie politique. *Revue de l'Écologie Politique* 6: 5–14.

Deléage, Jean-Paul. 1993c. Aux origines de la science écologique. *Revue de l'Écologie Politique* 7: 119–131.

Deléage, Jean-Paul. 1994a. Penser, agir. . . . *Revue de l'Écologie Politique* 9: 5–8.

Deléage, Jean-Paul. 1994b. L'écologie, science de l'homme et de son environnement. In *Les paradoxes de l'environnement*, ed. P. David. Albin Michel.

Deléage, Jean-Paul. 1994c. Eco-Marxist critique of political economy. In *Is Capitalism Sustainable?* ed. M. O'Connor. Guilford.

Deléage, Jean-Paul. 1997. Pour un écosocialisme. *Écologie et Politique* 20: 137–138.

Deléage, Jean-Paul, and Daniel Hémery. 1990. From ecological history to world ecology. In *The Silent Countdown*, ed. P. Brimblecombe. Springer-Verlag.

Deleuze, Gilles, and Félix Guattari. 1972. *Capitalisme et schizophrénie*, volume 1: *L'Anti-Oedipe.* Minuit.

Descartes, René. 1637. *Discourse on Method* (Bobbs-Merrill, 1950).

Descombes, Vincent. 1980. *Modern French Philosophy.* Cambridge University Press.

Devall, Bill, and George Sessions. 1985. *Deep Ecology: Living As If Nature Mattered.* Peregrine Smith.

di Norcia, Vincent. 1974–75. From critical theory to critical ecology. *Telos*, winter: 85–95.

Dobson, Andrew. 1995. *Green Political Thought*, second edition. Unwin Hyman.

Dorst, Jean. 1970. *La nature dé-naturée: pour une écologie politique.* Delachaux et Niestlé, Collection Points.

Drouin, Jean-Marc. 1991. *L'Écologie et son histoire.* Flammarion.

Dryzek, John. 1987. *Rational Ecology: Environment and Political Economy.* Blackwell.

Dryzek, John. 1994. Ecology and discursive democracy: Beyond liberal capitalism and the administrative state. In *Is Capitalism Sustainable?* ed. M. O'Connor. Guilford.

Dryzek, John. 1995. Political and ecological communication. *Environmental Politics* 4, no. 4: 13–30.

Dubos, René. 1972. *A God Within*. Scribner.

Duclos, Denis, ed. 1981. *De l'usine on peut voir la ville*. Échanges Sociales.

Duclos, Denis. 1989. *La Peur et le savoir: La société face à la science, la technique et leurs dangers*. La Découverte.

Duclos, Denis. 1991a. *Les Industriels et les risques pour l'environnement*. L'Harmattan.

Duclos, Denis. 1991b. *L'Homme face au risque technique*. L'Harmattan.

Duclos, Denis. 1993a. *De la civilité: Comment les sociétés apprivoisent la puissance*. La Découverte.

Duclos, Denis. 1993b. Les industrielles et l'environnement: Un nouveau paradigme? *Écologie Politique* 5, winter: 95–122.

Duclos, Denis. 1993c. Qu'attendons-nous de la nature? *Libération,* May 18.

Duclos, Denis. 1996. *Nature et démocratie des passions*. Presses Universitaires de France.

Duclos, Denis, and Jocelyne Smadja. 1985. Culture and public policy: The case of environmental policy in France. *Environmental Management* 9, no. 2: 135–140.

Dumont, René. 1973. *L'Utopie ou la mort!* Seuil. [*Utopia or Else*. Universe, 1975.]

Dumont, René. 1977. *Seule une écologie socialiste*. Laffont.

Dumont, René. 1978. Entretien. In *Pourquoi les écologistes font-ils de la politique?* Seuil.

Dumont, René. 1986. *Les Raisons de la colère*. Entente.

Dumont, René. 1988. *Un Monde intolérable: Le libéralisme en question*. Seuil.

Dumont, René. 1989. *Mes Combats*. Plon.

Dumont, René, with Charlotte Paquet. 1994. *Misère et chômage, libéralisme ou démocratie*. Seuil.

Duverger, Maurice. 1992. Naissance d'une écologie politique. *Le Monde*, March 19.

Eckersley, Robyn. 1992. *Environmentalism and Political Theory: Toward an Ecocentric Approach*. State University of New York Press.

Eckersley, Robyn. 1996. Greening liberal democracy: The rights discourse revisited. In *Democracy and Green Political Thought*, ed. B. Doherty and M. de Geus. Routledge.

Eckersley, Robyn. 1999. The discourse ethic and the problem of representing nature. *Environmental Politics* 8, no. 2: 24–49.

Ehrenfeld, David. 1978. *The Arrogance of Humanism*. Oxford University Press.

Eisendrath, Charles R. 1979. Environmentalism in France: The press forms a powerful "nebula." *Contemporary French Civilization* 3, no. 2: 212–225.

Ellul, Jacques. 1992. Si nous ne changeons pas de vie. . . . *Le Nouvel Observateur*, May 7–13: 27.

Evernden, Neil. 1992. *The Social Creation of Nature*. Johns Hopkins University Press.

Fabiani, Jean-Louis. 1986. Les Français et la protection de la nature. *Regards sur l'Actualité* 117, January: 31–38

Fages, J.B. 1980. *Comprendre Edgar Morin*. Privat.

Faivret, J.-P., J.-L. Missika, J.-L., and D. Wolton. 1980. *L'Illusion Écologique*. Seuil.

Falque, Max. 1986. Libéralisme et environnement. *Futuribles*, March: 40–54.

Faucher, Florence. 1999. *Les Habits Verts de la Politique*. Presses de la Fondation Nationale des Sciences Politiques.

Ferry, Luc. 1985. *French Philosophy of the Sixties* (University of Massachusetts Press, 1990).

Ferry, Luc. 1991–92. Entretien: Ce qui est dangéreux dans l'écologie. *Nouvel Observateur*, December 26–January 1: 30–31.

Ferry, Luc. 1992a. *Le Nouvel Ordre Écologique: L'arbre, l'animal, et l'homme*. Bernard Grasset. [*The New Ecological Order*, University of Chicago Press, 1995.]

Ferry, Luc. 1992b. Faut-il redouter l'écologie? *Le Point*, November 21.

Fischer, Frank, and Maarten Hajer, eds. 1999. *Living With Nature: Environmental Politics as Cultural Discourse*. Oxford University Press.

Foucault, Michel. 1976. *Histoire de la sexualité: La volonté de savoir*. Gallimard.

Foucault, Michel. 1980. *Power/Knowledge: Selected Interviews and Other Writings, 1972–1977*, ed. C. Gordon. Pantheon.

Frankel, Boris. 1987. *The Post-Industrial Utopians*. University of Wisconsin Press.

Frawley, Kevin. 1994. Evolving visions: Environmental management and nature conservation in Australia. In *Australian Environmental History*, ed. S. Dovers. Oxford University Press.

Gauchet, Marcel. 1990. Sous l'amour de la nature, la haine des hommes. *Le Débat* 60, May–August: 278–282.

Georgescu-Roegen, Nicholas. 1971. *The Entropy Law and the Economic Process*. Harvard University Press.

Gher, Sophie. 1992. L'itinéraire d'un économiste vert. *Le Monde*, March 22–23.

Goguel, François. 1989. Une vision simpliste des problèmes. *Le Figaro*, March 29.

Goodin, Robert E. 1992. *Green Political Theory*. Polity.

Gorz, André. 1964. *Stratégie ouvrière et néocapitalisme*. Seuil. [*Strategy for Labor: A Radical Proposal*. Beacon, 1967.]

Gorz, André. 1973. *Critique de la division du travail*. Seuil.

Gorz, André. 1975. *Écologie et politique*. (Seuil, 1978).

Gorz, André. 1977. *Écologie et liberté*. Galilée.

Gorz, André. 1980. *Adieu au prolétariat: Au delà du socialisme*. Galilée.

Gorz, André. 1988. *Métamorphoses du travail: Quête du sens. Critique de la raison économique*. Galilée.

Gorz, André. 1991. *Capitalisme, Socialisme, Écologie*. Galilée.

Gorz, André. 1993. Political ecology: Expertocracy or self-limitation. *New Left Review* 202: 55–67.

Goudie, Andrew. 1994. *The Human Impact on the Natural Environment*, fourth edition. MIT Press.

Gouget, Jean-Jacques. 1985. Environnement et productivisme: l'impossible alliance. *Reflets et Perspectives de la Vie Économique* 24, no. 4: 239–253

Grosser, Alfred. 1991. Ni Marx, ni Tarzan. *Libération*, October 2.

Guattari, Félix. 1989. *Les Trois Écologies*. Galilée. [*The Three Ecologies*. Athlone, 2000.]

Guattari, Félix. 1991a. La passion des machines: Un entretien avec Félix Guattari. *Terminal* 55: 44–45.

Guattari, Félix. 1991b. Qu'est-ce que l'écosophie? *Terminal* 56: 22–23.

Guattari, Félix. 1992. *Chaosmose*. Galilée. [*Chaosmosis; An Ethico-Aesthetic Paradigm*. Indiana University Press, 1995.]

Guillebaud, Jean-Claude. 1992. La pensée verte est-elle mûre? *Le Nouvel Observateur*, May 7–13: 19–20.

Guha, R., and J. Martinez-Alier. 1997. *Varieties of Environmentalism: Essays North and South*. Zed.

Gundersen, Adolf G. 1995. *The Environmental Promise of Democratic Deliberation*. University of Wisconsin Press.

Habermas, Jürgen. 1971. *Knowledge and Human Interests*. Beacon.

Habermas, Jürgen. 1975. *Legitimation Crisis*. Beacon.

Habermas, Jürgen. 1979. *Communication and the Evolution of Society*. Beacon.

Habermas, Jürgen. 1982. A reply to my critics. In *Habermas: Critical Debates*, ed. J. Thompson and D. Held. MIT Press.

Habermas, Jürgen. 1984. *The Theory of Communicative Action*, volume 1: *Reason and the Rationalization of Society*. Beacon.

Habermas, Jürgen. 1987b. *The Theory of Communicative Action*, volume 2: *Lifeworld and System: A Critique of Functionalist Reason*. Beacon.

Habermas, Jürgen. 1990. Discourse ethics: Notes on a program of philosophical justification. In *Moral Consciousness and Communicative Action*. MIT Press.

Harari, Josué V., and David F. Bell, eds. 1982. *Michel Serres, Hermes: Literature, Science, Philosophy*. Johns Hopkins University Press.

Harbers, Hans. 1995. Review of Latour, *We Have Never Been Modern*. *Science, Technology and Human Values* 20, no. 2: 270–275.

Hardin, Garrett. 1968. "The Tragedy of the Commons." *Science* 162: 1243–1248.

Hastings, Michel. 1992. Le discours écologiste: une utopie syncrétique. *Regards sur l'actualité* 178, February: 17–29.

Hazelrigg, Lawrence. 1995. *Cultures of Nature: An Essay on the Production of Nature*. University Press of Florida.

Hayden, Patrick. 1997. Gilles Deleuze and naturalism: A convergence with ecological theory and politics. *Environmental Ethics* 19, no. 2: 185–205.

Hayward, Tim. 1994. *Ecological Thought: An Introduction*. Polity.

Hayward, Tim. 1998. *Political Theory and Ecological Values*. St. Martin's.

Heim, Roger. 1952. *Destruction et protection de la nature*. Librairie A. Colin.

Hughes, H. Stuart. 1966. *The Obstructed Path: French Social Thought in the Years of Desperation 1930–1960*. Harper & Row.

Jacob, François. 1970. *La Logique du vivant*. Gallimard.

Jacob, Jean. 1995. *Les sources de l'écologie politique*. Panoramiques-Corlet.

Jacob, Jean. 1999. *Histoire de l'écologie politique*. Albin Michel.

Johnson, Lawrence E. 1991. *A Morally Deep World: An Essay on Moral Significance and Environmental Ethics*. Cambridge University Press.

Journès, Claude. 1979. Les idées politiques du mouvement écologique. *Revue Française de Science Politique* 2, no. 39: 230–254.

Journès, Claude. 1984. Idées économiques et sociales des écologistes. *Projet* 182: 215–223.

Juquin, Pierre, Carlos Antunes, Penny Kemp, Isabelle Stengers, Wilfred Telkamper, and Frieder Otto Wolf. 1990. *Pour une alternative verte en Europe*. La Découverte.

Kalaora, Bernard. 1998. *Au-delà de la nature, l'environnement: L'observation sociale de l'environnment*. L'Harmattan.

Katz, Eric. 1997. *Nature as Subject: Human Obligation and Natural Community*. Rowman & Littlefield.

Kempf, Hervé. 1994. *La Baleine qui cache la forêt: Enquêtes sur les pièges de l'écologie*. La Découverte.

Keulartz, Jozef. 1999. Engineering the environment: The politics of "nature development." In *Living With Nature*, ed. F. Fischer and M. Hajer. Oxford University Press.

Kitschelt, Herbert P. 1985. Review of Capra and Spretnak, *The Global Promise of Green Politics*. *Theory and Society* 14, no. 4: 525–533.

Kitschelt, Herbert. 1990. La gauche libertaire et les écologistes français. *Revue Française de Science Politique* 40, no. 3: 339–365.

Kitsching, R. L. 1983. *Systems Ecology: An Introduction to Ecological Modelling*. University of Queensland Press.

Kofman, Myron. 1996. *Edgar Morin: From Big Brother to Fraternity*. Pluto.

Kölher, Jochen. 1983. Serge Moscovici: Versuch über die menschliche Geschichte der Natur. *Philosophische Rundschau* 30: 23–30.

Laborit, Henri. 1968. *Biologie et structure*. Gallimard.

Laborit, Henri. 1974. *La Nouvelle Grille*. Laffont.

Lacan, Jacques. 1977. *Écrits: A Selection*. Norton.

Lacan, Jacques. 1988. *The Seminar of Jacques Lacan, Book I: Freud's Papers on Technique*. Norton.

Lalonde, Brice, and Dominique Simmonet. 1978. *Quand vous voudrez*. Jean-Jacques Pauvert.

Lalonde, Brice. 1981. *Sur la vague verte*. Laffont.

Lalonde, Brice. 1993. Les origines du mouvement écologiste. In *Le Défi écologiste*, ed. M. Abélès. L'Harmattan.

Larrère, Catherine. 1993. Éthique et environnement: À propos du Contrat Naturel. *Écologie Politique* 5, winter: 27–49.

Larrère, Catherine. 1997. *Les philosophies de l'environnement*. Presses Universitaires de France.

Larrère, Catherine, and Raphaël Larrère. 1997. *Du bon usage de la nature: Pour une philosophie de l'environnement*. Aubier.

Lascoumes, Pierre. 1991. Le Droit de l'environnement. In *Environnement et Gestion de la planète*, ed. J. Theys (*La Documentation Française* no. 250, March–April: 61–65).

Lascoumes, Pierre. 1994. *L'Écopouvoir: environnements et politiques*. La Découverte.

Lascoumes, Pierre. 1998. "La scène publique: nouveau passage obligé des décisions?" *Annales des Mines* no. 10: 51–62.

Lascoumes, Pierre. 1999. "Productivité des controverses et renouveau de l'expertise." *Les Cahiers de la sécurité intérieure* 38, "fourth trimester": 75–96.

Latour, Bruno. 1984. *Les microbes: Guerre et paix*. Métailié.

Latour, Bruno. 1987a. Enlightenment without the critique: A word on Michel Serres' philosophy. In *Contemporary French Philosophy*, ed. A. Phillips Griffiths. Cambridge University Press.

Latour, Bruno. 1987b. *Science in Action: How to Follow Scientists and Engineers through Society*. Harvard University Press.

Latour, Bruno. 1990. Postmodern? No, simply amodern! Steps towards an anthropology of science. *Studies in History and Philosophy of Science* 21, no. 1: 145–171.

Latour, Bruno. 1991. *Nous n'avons jamais été modernes*. La Découverte.

Latour, Bruno. 1993. Arrachement ou attachement à la nature? *Écologie Politique* 5, winter: 15–26.

Latour, Bruno. 1994a. Pragmatogonies: A mythical account of how humans and nonhumans swap properties. *American Behavioral Scientist* 37, no. 6: 791–808.

Latour, Bruno. 1994b. Esquisse d'un parlement des choses. *Écologie Politique* 10: 97–115.

Latour, Bruno. 1999a. *Pandora's Hope: Essays on the Reality of Science Studies.* Harvard University Press.

Latour, Bruno. 1999b. *Politiques de la nature: Comment faire entrer les sciences en démocratie.* La Découverte.

Latour, Bruno, and Steve Woolgar. 1979. *Laboratory Life: The Social Construction of Scientific Facts.* Sage.

Latour, Bruno, Cécile Schwartz, and Florian Charvolin. 1991. Crises des environnements: Défis aux sciences humaines. *Futur antériur* 6: 28–56.

Launay, Stephen. 1997. Une génèse de la conscience écologique: La pensée de Bertrand Jouvenel. *Écologie et Politique* 21, autumn–winter: 101–123.

Laurens, André. 1991. L'Écologie comme humanisme. *Le Monde*, January 6–7.

Lavely, John. 1967. Personalism. In *Encyclopedia of Philosophy.* Collier Macmillan.

Lebreton, Philippe. 1977. *L'Éco-logique.* Interéditions.

Lebreton, Philippe. 1978. *L'Ex-croissance: Les chemins de l'écologisme.* Denoël.

Lecourt, Dominique. 1997. *L'avenir du progrès.* Textuel.

Leger, Danièle, and Bertrand Hervieu. 1979. *Le Retour à la nature: Au fond de la forêt . . . l'État.* Seuil.

Legoff, Jean-Pierre. 1981. De Mai 68 à l'écologie politique: Point de vue sur un itinéraire. *Que Faire Aujourd'hui?* 16, October: 4–7.

Lenoir, Yves. 1992. *La vérité sur l'effet de serre: Le dossier d'une manipulation planétaire.* La Découverte.

Leopold, Aldo. 1949. *A Sand County Almanac* (Ballantine, 1970).

Les Verts. 1994. *Le Livre des Verts: Dictionnaire de l'Écologie Politique.* Félin.

Levi, Albert William. 1969. *Humanism and Politics: Studies in the Relationship of Power and Value in the Western Tradition.* Indiana University Press.

Light, Andrew, and Eric Katz, eds. 1996. *Environmental Pragmatism.* Routledge.

Lilla, Mark, ed. 1994. *New French Thought: Political Philosophy.* Princeton University Press.

Lipietz, Alain. 1987. Rebel sons: The regulation school. *French Politics and Society* 5: 17–26.

Lipietz, Alain. 1988a. Building an alternative movement in France. *Rethinking Marxism* 1, no. 3: 80–99.

Lipietz, Alain. 1988b. Reflections on a tale: The Marxist foundations of the concepts of regulation and accumulation. *Studies in Political Economy* 26, summer: 7–44.

Lipietz, Alain. 1989. *Choisir l'audace: Une alternative pour le XXIe siècle.* La Découverte. [*Towards a New Economic Order: Postfordism, Ecology and Democracy.* Oxford University Press, 1992.]

Lipietz, Alain. 1992a. A regulationist approach to the future of urban ecology. *Capitalism, Nature, Socialism* 3, no. 3: 101–110.

Lipietz, Alain. 1992b. *Berlin, Bagdad, Rio*. Quai Voltaire.

Lipietz, Alain. 1993a. *Vert Espérance: L'Avenir de l'écologie politique*. La Découverte. [*Green Hopes: The Future of Political Ecology*. Blackwell, 1995.]

Lipietz, Alain (with Gérard Bramoullé). 1993b. Face à face: Libéralisme ou écologisme. *Krisis* 15, September: 37–50.

Lipietz, Alain. 1994. De la régulation aux conventions: Le grand bond en arrière? *Actuel Marx* 15: 39–48.

Lipietz, Alain. 1999. *Qu'est-ce que l'écologie politique? La grande transformation du XXIᵉ siècle*. La Découverte.

Lipietz, Alain. 2000. Political ecology and the future of Marxism. *Capitalism, Nature, Socialism* 11, no. 1: 69–85.

Little, Adrian. 1996. *The Political Thought of André Gorz*. Routledge.

Loubet del Bayle, Jean-Louis. 1987. *Les non-conformistes des années trente. Une tentative de renouvellement de la pensée politique française*. Seuil.

Lovelock, James. 1995. *Gaia: A New Look at Life on Earth*, revised edition. Oxford University Press.

Luke, Timothy. 1997. *Ecocritique: Contesting the Politics of Nature, Economy, and Culture*. University of Minnesota Press.

Lukes, Steven. 1991. *Moral Conflict and Politics*. Clarendon.

Lyotard, Jean-François. 1979. *La Condition postmoderne*. Minuit.

Macnaghten, Phil, and John Urry. 1998. *Contested Natures*. Sage.

Marx, Karl. 1867. *Das Kapital*, volume I. (Ullstein, 1969).

Martell, Luke. 1994. *Ecology and Society: An Introduction*. University of Massachusetts Press.

Mathews, Freya. 1991. *The Ecological Self*. Routledge.

Matthews, Eric. 1996. *Twentieth-Century French Philosophy*. Oxford University Press.

McCarthy, Thomas. 1978. *The Critical Theory of Jürgen Habermas*. MIT Press.

McCloskey, H. J. 1983. *Ecological Ethics and Politics*. Rowman & Littlefield.

Meadows, Donella H., Dennis Meadows, Jørgen Randers, and Jørgen Behrens III. 1972. *The Limits to Growth*. Universe.

Mills, Mike. 1996. Green democracy: The search for an ethical solution. In *Democracy and Green Political Thought*, ed. B. Doherty and M. de Geus. Routledge.

Mongin, Olivier. 1994. *Face au scepticisme: Les mutations du paysage intellectuel ou l'invention de l'intellectuel démocratique*. La Découverte.

Montaigne, Michel de. 1967. *Oeuvres complètes*. Seuil.

Morin, Edgar. 1951. *L'Homme et la mort* (Seuil, 1970).

Morin, Edgar. 1973. *Le Paradigme perdu: la nature humaine.* Seuil.

Morin, Edgar. 1977. *La méthode, I: La nature de la nature.* Seuil. [*The Nature of Nature.* Lang, 1992.]

Morin, Edgar. 1980. *La méthode, II: La vie de la vie.* Seuil.

Morin, Edgar. 1990. Messie, mais non. In *Arguments pour une méthode (Autour d'Edgar Morin)*, ed. D. Bougnoux et al. Seuil.

Morin, Edgar. 1992. Pour une pensée écologisée. In *La terre outragée*, ed. J. Theys and B. Kalaora. Diderot.

Morin, Edgar, and Anne Brigitte Kern. 1993. *Terre-Patrie.* Seuil. [*Homeland Earth: A Manifesto for the New Millenium.* Hampton, 1998.]

Mortley, Raoul. 1991. *French Philosophers in Conversation.* Routledge.

Moscovici, Serge. 1968. *Essai sur l'histoire humaine de la nature.* (Flammarion, 1977).

Moscovici, Serge. 1972. *La Société contre nature.* Seuil, 1994. [*Society against Nature: The Emergence of Human Societies.* Harvester, 1976.]

Moscovici, Serge. 1974. *Hommes Domestiques et hommes sauvages* (Christian Bourgeois, 1979).

Moscovici, Serge. 1976. Die Widerverzauberung der Welt. In *Jenseits der Krise*, ed. A. Touraine. Syndikat.

Moscovici, Serge. 1978. Entretien. In *Pourquoi les écologistes font-ils de la politique?* Seuil.

Moscovici, Serge. 1990. Questions for the twenty-first century. *Theory, Culture & Society* 7, no. 4: 1–19.

Moscovici, Serge. 1993. La polymérisation de l'écologie. In *Le Défi écologiste*, ed. M. Abélès. L'Harmattan.

Moscovici, Serge. 1994. Postface à l'edition de 1994. In *La société contre nature.* Seuil.

Müller-Rommel, Ferdinand. 1985. The Greens in Western Europe: Similar but different. *International Political Science Review* 6, no. 4: 483–499.

Mumford, Lewis. 1938. *The Culture of Cities.* Harcourt.

Naess, Arne, and David Rothenberg. 1989. *Ecology, Community and Lifestyle.* Cambridge University Press.

Nash, Roderick. 1982. *Wilderness and the American Mind*, third edition. Yale University Press.

Nick, Christophe. 1991. Les écolos fachos. *Actuel* 10, October: 10–20.

Norton, Bryan. 1984. Environmental ethics and weak anthropocentrism. *Environmental Ethics* 6: 131–148.

Norton, Bryan. 1991. *Toward Unity among Environmentalists.* Oxford University Press.

O'Connor, James. 1998. *Natural Causes: Essays in Ecological Marxism.* Guilford.

Odum, Eugene P. 1974. Environmental ethics and the attitude revolution. In *Philosophy and the Environmental Crisis*, ed. W. Blackstone. University of Georgia Press.

O'Neill, John. 1993. *Ecology, Policy and Politics: Human Well-Being and the Natural World.* Routledge.

Onfray, Michel. 1992. Les babas-cool du Maréchal. *Le Nouvel Observateur*, May 7–13: 20–21.

Ophuls, William, and A. Stephen Boyan Jr. 1992. *Ecology and the Politics of Scarcity Revisited: The Unraveling of the American Dream.* Freeman.

Ottman, Henning. 1982. Cognitive interests and self-reflection. In *Habermas: Critical Debates*, ed. J. Thompson and D. Held. MIT Press.

Oudin, Bernard. 1996. *Pour en finir avec les écolos.* Gallimard.

Paehlke, Ronald. 1989. *Environmentalism and the Future of Progressive Politics.* Yale University Press.

Paraire, Philippe. 1992. *L'Utopie verte: Écologie des riches, écologie des pauvres.* Hachette.

Parodi, Jean-Luc. 1979. Essai de problématique du mouvement écologiste: les écologistes et la tentation politique. *Revue Politique et Parlementaire* 81, no. 878: 25–43.

Pascal, Blaise. 1971. *Pensées et Opuscules.* Hachette.

Passet, René. 1979. *L'Économique et le vivant.* Payot.

Passet, René. 1989. Que l'économie serve la biosphère. *Le Monde Diplomatique* August: 4–5.

Passmore, John. 1974. *Man's Responsibility Toward Nature: Ecological Problems and Western Tradition.* Scribner.

Pelletier, Philippe. 1993. *L'Imposture écologiste.* GIP Reclus.

Pelt, Jean-Marie. 1977. *L'Homme re-naturé.* Seuil.

Pelt, Jean Marie. 1980. *Les plantes, leurs amours, leurs problèmes, leurs civilisations.* Fayard.

Pelt, Jean-Marie. 1990. *Le Tour du monde d'un écologiste.* Fayard.

Pelt, Jean-Marie. 1993. *Une Leçon de nature.* L'Esprit du Temps.

Pelt, Jean-Marie. 1995. *Dieu de l'univers: Science et foi.* Fayard.

Pelt, Jean-Marie. 1996. *De l'univers à l'être: réflexions sur l'évolution.* Fayard.

Pelt, Jean-Marie. 2000. *Plantes et aliments transgéniques.* Fayard.

Pepper, David. 1984. *The Roots of Modern Environmentalism* (Routledge, 1989).

Pepper, David. 1993. *Eco-Socialism: From Deep Ecology to Social Justice.* Routledge.

Pierce, Roy. 1966. *Contemporary French Political Thought.* Oxford University Press.

Pinchemel, Philippe. 1980. *France: A Geographical, Social, and Economic Survey* (Cambridge University Press, 1987).

Ponting, Clive. 1991. *A Green History of the World: The Environment and the Collapse of Great Civilizations.* Penguin.

Postel-Vinay, Olivier. 1998. Ecologisme versus libéralisme. *Sociétale* 23, December: 70–74.

Plumwood, Val. 1995. Has democracy failed ecology? An ecofeminist perspective. *Environmental Politics* 4, no. 4: 134–168.

Prendiville, Brendan. 1993. *L'Écologie, la Politique autrement? Culture, sociologie et histoire des écologistes.* L'Harmattan.

Press, Daniel. 1994. *Democratic Dilemmas in the Age of Ecology: Trees and Toxics in the American West.* Duke University Press.

Pronier, Raymond, and Vincent Jacques le Seigneur. 1992. *Génération Verte: Les écologistes en politique.* Presses de la Renaissance.

Rauch, R. William Jr. 1972. *Politics and Belief in Contemporary France: Emmanuel Mounier and Christian Democracy, 1932–1950.* Martinus Nijhoff.

Robin, Jacques. 1975. *De la croissance économique au développement humain.* Seuil.

Robin, Jacques. 1989a. Le choix écologique contre la volonté de puissance. *Le Monde Diplomatique,* July: 16–17.

Robin, Jacques. 1989b. *Changer d'ère.* Seuil.

Robin, Jacques. 1990. Edgar Morin et le "groupe des Dix." In *Arguments pour une méthode (Autour d'Edgar Morin),* ed. D. Bougnoux et al. Seuil.

Rolston, Holmes III. 1988. *Environmental Ethics: Duties to and Values in the Natural World.* Temple University Press.

Roose, Frank de, and Philippe Van Parijs. 1991. *La pensée écologiste: Essai d'inventaire à l'usage de ceux qui la pratiquent comme ceux qui la craignent.* De Boeck Université.

Rosnay, Joël de. 1975. *Le Macroscope: Vers une vision global.* Seuil. [*Macroscope: A New World Scientific System.* Harper & Row, 1979.]

Rosnay, Joël de. 1991. L'avenir de l'écologie: une pensée locale, un geste global. In *Problèmes politiques et sociaux,* no. 651: *La sensibilité écologique en France,* ed. P. Alphandéry et al. La Documentation Française.

Rosnay, Joël de. 1995. *L'Homme symbiotique: Regards sur le troisième millénaire.* Seuil. [*Symbiotic Man: A New Understanding of Life and a Vision of the Future.* McGraw-Hill, 2000.]

Rougement, Denis de. 1972. *Les Dirigeants et les finalités de la société occidentale.* Centre de Recherches Européennes.

Rougement, Denis de. 1977. *L'Avenir est notre affaire.* Stock. [*The Future Is Within Us.* Pergamon, 1983.]

Roy, Christian. 1997. Entre pensée et nature: le personnalisme gascon. In *Bernard Charbonneau: Une vie entière à dénoncer la grande imposture*, ed. J. Prades. Erès.

Rudolf, Florence. 1998. *L'Environnement, une construction sociale: pratiques et discours sur l'environnement en Allemagne et en France*. Presses Universitaires de Strasbourg.

Ryle, M. 1988. *Ecology and Socialism*. Rodins.

Sagoff, Mark. 1988. *The Economy of the Earth*. Cambridge University Press.

Sainteny, Guillaume. 1991. *Les Verts*. Presses Universitaires de France.

Sainteny, Guillaume. 1993. L'Écologisme en Allemagne et en France. *Écologie Politique* 6, spring: 15–28.

Sainteny, Guillaume. 2000. *L'introuvable écologisme français?* Presses Universitaires de France.

Saint Marc, Philippe. 1971. *Socialisation de la nature*. Stock.

Saint Marc, Philippe. 1978. *Progrès ou déclin de l'homme?* Stock.

Saint Marc, Philippe. 1994. *L'Économie barbare*. Frison-Roche.

Saint-Ouen, François. 1995. *Denis de Rougement: Introduction à sa vie et son oeuvre*. Georg.

Saint-Upéry, Marc. 1992. Faut-il avoir peur de Luc Ferry? *Revue de l'Écologie Politique* 2, spring 141–147.

Sale, Kirkpatrick. 1991. *Dwellers in the Land*. New Society.

Sandoval, C. A. 1995. Michel Serres' Philosophy of the Educated Third. *Philosophy Today* 39, no. 2: 107–118.

Sartori, Giovanni. 1987. *The Theory of Democracy Revisited*. Chatham House.

Saward, Michael. 1993. Green Democracy? In *The Politics of Nature: Explorations in Green Political Theory*, ed. A. Dobson and P. Lucardie. Routledge.

Saward, Michael. 1996. Must Democrats Be Environmentalists? In *Democracy and Green Political Thought*, ed. B. Doherty and M. de Geus. Routledge.

Schaffer, Simon. 1991. The Eighteenth Brumaire of Bruno Latour. *Studies in History and Philosophy of Science* 22, no. 1: 174–192.

Schneider, Marcel. 1978. *Jean-Jacques Rousseau et l'espoir écologiste*. Pygmalion.

Serres, Michel. 1972. *Hermès II: L'Interférence*. Minuit.

Serres, Michel. 1975. *Feux et signaux de brume: Zola*. Grasset.

Serres, Michel. 1977. *La naissance de la physique dans le texte de Lucrèce. Fleuves et turbulences*. Minuit.

Serres, Michel. 1982. *Hermes: Literature, Science, Philosophy*, ed. J. Harari and D. Bell. Johns Hopkins University Press.

Serres, Michel. 1990. *Le Contrat naturel*. François Bourin. [*The Natural Contract*. University of Michigan Press, 1995.]

Serres, Michel. 1991. *Le Tiers-Instruit*. François Bourin. [*The Troubador of Knowledge*. University of Michigan Press, 1997.]

Serres, Michel. 1992a *Éclaircissements: Cinq Entretiens avec Bruno Latour.* François Bourin.

Serres, Michel. 1992b. Un entretien avec Michel Serres: "Nous entrons dans une période où la morale devient objective." *Le Monde*, January 21.

Serres, Michel. 1992c. A mes contemporains, ces hémiplégiques: Un entretien avec le philosophe du 'Contrat Naturel.' *Le Nouvel Observateur*, May 7–13: 24–25.

Serres, Michel. 1993. Le Contrat naturel: Un entretien avec Michel Serres. *Krisis* 15, September: 105–112.

Serres, Michel. 2000. *Retour au Contrat naturel.* Bibliothèque Nationale de France.

Shull, Tad. 1999. *Redefining Red and Green: Ideology and Strategy in European Political Ecology.* State University of New York Press.

Simmonet, Dominique. 1982. *L'Écologisme.* Presses Universitaires de France.

Sintomer, Yves. 1991–92. Review of Gorz, *Capitalisme, Socialisme, Écologie. Revue de l'Écologie Politique* 1, winter: 107–111.

Soper, Kate. 1995. *What Is Nature? Culture, Politics and the Non-Human.* Blackwell.

Snyder, Gary. 1974. *Turtle Island.* New Directions.

Stevens, Anne. 1991. Culture and public policy: The case of environmental policy in France. *Modern and Contemporary France* 44, January: 25–33.

Taylor, Bob Pepperman. 1992. *Our Limits Transgressed: Environmental Political Thought in America.* University of Kansas Press.

Taylor, Paul W. 1986. *Respect for Nature: A Theory of Environmental Ethics.* Princeton University Press.

Tazieff, Haroun. 1989. Non à un parti politique vert. *Le Figaro*, March 29.

Tazieff, Haroun. 1990. L'Homme et la nature. *Philosophie Politique* 6: 17–31.

Tazieff, Haroun. 1992. *La Terre va-t-elle cesser de tourner? Pollutions réelles, pollutions imaginaires.* Segners.

Theys, Jacques. 1993. Le savant, le technicien, et le politique. In *La Nature en politique, ou l'enjeu philosophique de l'écologie*, ed. D. Bourg. L'Harmattan.

Thomas, Bernard. 1992. *Lettre ouverte aux écologistes qui nous pompent l'air.* Albin Michel.

Todorov, Tzvetan. 1998. *Le jardin imparfait: La pensée humaniste en France.* Grasset.

Torgerson, Douglas. 1999. *The Promise of Green Politics: Environmentalism and the Public Sphere.* Duke University Press.

Touraine, Alain. 1983. *Anti-Nuclear Protest: The Opposition to Nuclear Energy in France.* Cambridge University Press.

Trom, Dany. 1990. Le parler vert: Réflexions sur les structures discursives de l'idéologie écologiste. *Politix* 9: 44–52.

Troude-Chastenet, Patrick. 1998. Jacques Ellul, précurseur de l'écologie politique. *Écologie et Politique* 22, spring: 105–122.

Vadrot, Claude-Marie. 1978. *L'Écologie, histoire d'une subversion*. Syros.

Van Wyck, Peter. 1997. *Primitives in the Wilderness: Deep Ecology and the Missing Human Subject*. State University of New York.

Viard, Jean. 1990. *Le Tiers Espace: Essai sur la nature*. Méridiens Klincksieck.

Villalba, Bruno. 1997. La génèse inachevée des Verts. *Vingtième Siècle* 53, January–March: 85–89.

Virilio, Paul. 1978. *Défense populaire et luttes écologiques*. Galilée.

Vogel, Steven. 1995. New Science, New Nature: The Habermas-Marcuse Debate Revisited. In *Technology and the Politics of Knowledge*, ed. A. Feenberg and A. Hannay. Indiana University Press.

Waechter, Antoine. 1990. *Dessine-moi une planète*. Albin Michel.

Weber, Max. 1949. *The Methodology of the Social Sciences*, ed. E. A. Shils and H. Finch. Free Press.

Wells, David, and Tony Lynch. 2000. *The Political Ecologist*. Ashgate.

Wesling, Donald. 1997. Michel Serres, Bruno Latour, and the Edges of Historical Periods. *Clio* 2, no. 2: 189–204.

Weston, Anthony. 1996. Beyond Intrinsic Value: Pragmatism in Environmental Ethics. In *Environmental Pragmatism*, ed. A. Light and E. Katz. Routledge.

White, Daniel R. 1998. *Postmodern Ecology: Communication, Evolution and Play*. State University of New York Press.

White, Stephen K. 1988. *The Recent Work of Jürgen Habermas: Reason, Justice and Modernity*. Cambridge University Press.

Whitebook, Joel. 1996. The Problem of Nature in Habermas. In *Minding Nature*, ed. D. Macauley. Guilford.

Whiteside, Kerry. 1988. *Merleau-Ponty and the Foundation of an Existential Politics*. Princeton University Press.

Whiteside, Kerry. 1992. The Political Practice of the Verts. *Modern and Contemporary France* 48, January: 14–21.

Whiteside, Kerry. 1994. Hannah Arendt and Ecological Politics. *Environmental Ethics* 16, winter: 339–358.

Whiteside, Kerry. 1995. The Resurgence of Ecological Political Thought in France. *French Politics and Society* 13, no. 3: 1–16.

Whiteside, Kerry. 1996. Regulation, Ecology, Ethics: The Red-Green Politics of Alain Lipietz, *Capitalism, Nature, Socialism* 7, no. 3: 31–55

Whiteside, Kerry H. 1997a. René Dumont and the Fate of Political Ecology in France. *Contemporary French Civilization* 21, no. 1: 1–17.

Whiteside, Kerry. 1997b. French Ecosocialism: From Utopia to Contract. *Environmental Values* 6, no. 3: 99–124.

Whiteside, Kerry H. 1999. Justice Uncertain: Judith Shklar on Liberalism, Skepticism and Equality. *Polity* 31, no. 3: 501–524.

Williams, Bernard. 1980. Political Philosophy and the Analytical Tradition. In *Political Theory and Political Education*, ed. M. Richter. Princeton University Press.

Williams, Raymond. 1995. Socialism and Ecology. *Capitalism, Nature, Socialism* 6, no. 1: 41–57.

Winner, Langdon. 1986. *The Whale and the Reactor: A Search for Limits in an Age of High Technology*. University of Chicago Press.

Wissenburg, Marcel. 1998. *Green Liberalism: The Free and the Green Society*. University College of London Press.

Index